in Translation?

Bioscience—Lost in Translation?

How precision medicine closes the innovation gap

Professor Richard Barker

Centre for the Advancement of Sustainable Medical
Innovation (CASMI),
Oxford, UK

OXFORD
UNIVERSITY PRESS

OXFORD

UNIVERSITY PRESS

Great Clarendon Street, Oxford, OX2 6DP,
United Kingdom

Oxford University Press is a department of the University of Oxford.
It furthers the University's objective of excellence in research, scholarship,
and education by publishing worldwide. Oxford is a registered trade mark of
Oxford University Press in the UK and in certain other countries

First Edition published in 2016

Impression: 1

Published in the United States of America by Oxford University Press
198 Madison Avenue, New York, NY 10016, United States of America

British Library Cataloguing in Publication Data
Data available

Library of Congress Control Number: 2016943988

ISBN 978-0-19-873778-0

Printed and bound by
CPI Group (UK) Ltd, Croydon, CR0 4YY

Foreword

It's my pleasure to commend this timely and important book from Richard Barker, Chair of the Precision Medicine Catapult and one of the UK's leading figures in precision medicine.

We are seeing unprecedented breakthroughs here in the UK and across the world in biomedicine and health technology that have major potential to save or enhance patients' lives as never before.

At the basic science level these are extraordinary times with a range of genuinely transformational new treatments and cures in the pipeline. From cancer to rare diseases, and to major chronic illnesses that bring such significant costs to individuals and society.

But the pace, scale and nature of these transformations challenge the traditional siloed models of both long-chain R+D, and the assessment, adoption and reimbursement of new treatments. The quiet revolution of patient centered research and Precision Medicine drives and demands a much closer integration of treatment and research.

Having launched the UK's Life Sciences Strategy with the Prime Minister in 2011, I am personally and deeply committed to bringing in the changes we need, both nationally and internationally. Here in the UK I have launched the Early Access to Medicines Scheme and the Accelerated Access Review for a new model of NHS partnership in the innovation department, to drive forward crucial reforms to the UK landscape.

The thesis of this book is that the new era of precision medicine holds the most important key to revolutionising biomedical science and 21st century healthcare. I agree. So I am delighted to commend this provocative yet practical book to all who want to see advances in bioscience benefit patients more rapidly and more effectively.

George Freeman MP,
Minister for Life Sciences

Preface and acknowledgements

This book is written to focus the minds of medical professionals, life science researchers, policy-makers, and industry executives on a problem that frustrates so many of their efforts to advance medical innovation. Much too much promising bioscience has been—and continues to be—lost in translation. Over more than 25 years of observing, managing, and studying innovation in biomedicine, I have become convinced we can and must overcome these gaps in translation, if we set out to understand their origins.

Other domains of human endeavour, especially the world of information technology, have seen dramatic practical advances over this same time period: products with better performance, at lower cost, reaching the consumer faster each year. In contrast, in medical innovation we have seen huge investments yield fewer products, slower and at greater cost, year on year. The result, of course, is a process that is unsustainable. We all need this situation to change radically, both in our professional roles and as patients and tax-paying citizens. I have written this book as my contribution to driving the changes we need to close the innovation gap.

I have been privileged to work in a wide variety of healthcare organizations on both sides of the Atlantic. They have included universities, private companies, public sector bodies, and the voluntary sector. I see the innovation challenge as a call to teamwork across all of them.

The groundwork for my thinking was laid in leadership roles at IBM, Chiron, iKnowMed, Molecular Staging, and the Association of the British Pharmaceutical Industry (ABPI). While working on medical innovation issues in each of them I have benefited greatly from the insights of colleagues far too numerous to mention. But I would like to pay particular tribute to a few who have particularly influenced my recent thinking—to my senior Centre for the Advancement of Sustainable Medical Innovation (CASMI) colleagues John Bell and Hugh Watkins at Oxford, John Tooke and Rob Horne at UCL, and to a very stimulating group of (largely) young CASMI members and associates, notably Jack Scannell, David Brindley, Natasha Davie, Liz Morrell, Megan Morys, Elin Haf Davies, Sarah Garner, Rosie Pigott, Stuart Faulkner, Suzanne Ii, and my CASMI co-founder, Nick Edwards.

I have drawn freely on work that CASMI has done on areas such as open innovation, new adaptive models of development, and the spread and adoption of new technology. This work has been supported by the Wellcome Trust, the Oxford Biomedical Research Centre, the UK government, the European Innovative Medicines Initiative, and Nesta. I'm grateful for their support.

I have also learnt many useful lessons from colleagues at the Health Innovation Network as they seek to spread valuable innovations across South London, including Chris Streather, Tara Donnelly, Zoe Lelliott, Anna King and Guy Boersma. Precision medicine is a major theme of this book, and I've gained insights from the team starting up the UK Precision Medicine Catapult. I have the honour to chair both these innovation enterprises; also Image Analysis, led by Olga Kubassova, a digital health company that is transforming the productivity of MRI. Other inspirational individuals further afield include Aubrey de Grey, Mike Cope, Liz Parrish, and Avi Roy, all of whom are dedicated to tackling the process and diseases of aging.

My particular thanks to Jack Scannell, Anoop Maini, and Avi Roy for reading and commenting on all or part of the book.

I have also had the privilege of observing great innovators in the pharmaceutical industry, in recent years as a board member of Celgene. They have convinced me that speed, open-mindedness, and creativity are quite compatible with the challenging, long-term business of bringing new drugs to patients.

In contrast to much popular supposition, regulators are largely supporting change rather than resisting it. Among current and former regulators, I'd highlight Hans-Georg Eichler, Thomas Longgren, and Tomas Salmonson at the European Medicines Agency, and Janet Woodcock and Mac Lumpkin (now with the Gates Foundation) at the US Food & Drug Administration. Successive leaders at the Medicines and Healthcare Products Regulatory Agency, including Alasdair Breckenridge, Kent Woods, Ian Hudson, Gordon Duff, and Mike Rawlins have helped in pioneering thinking that is reflected in this book. And many progressive life science industry leaders are partners in change.

I would like to particularly acknowledge the leadership of David Cooksey, whose seminal 2006 report paved the way for many efforts in this area, including my own. He has been an unstinting force behind creating a future in which life sciences make greater practical impact on our lives, faster and more affordably.

This is a book about the pressing need for change, to make the medical innovation process sustainable. It does not set out to explain in detail how things are done today: I will refer readers to some useful sources for this background. Instead it explores where and why current models are unsustainable and what we can do about it.

I have deliberately sought to look at the innovation process 'from soup to nuts'— from the initial idea of a product to the final benefit felt by the patient. I hope the book will be read by experts in the various stages of this long process, but I ask their understanding where—in my aim of encompassing it all—I don't do justice to the details that they know better than me. The references I include are sometimes to specific original research but also to reviews or even popular accounts where they seem accurate.

I have sought to take a sector-wide perspective, including pharmaceuticals, diagnostics, medical devices, and digital health. These are converging in this emerging era of precision medicine. However, much of the 'gap analysis' and many of the prescriptions for closing the gaps focus on pharmaceuticals. This is for two reasons: there has been much more public discussion and debate about drugs than about the other life science industry areas; and I believe there are genuinely more problems to be solved. Some of the solutions, though, apply across the sector and I will try to draw this out.

Despite its analysis of 'gaps in translation', this is fundamentally an optimistic book. I see many signs that the changes we need are starting to be made, but inertia is strong, so clear sight and determination are critical. I hope to reinforce both.

As always, I am immensely grateful for the support and patience of my wife, Michaela. I also greatly appreciate the enthusiastic sponsorship and guidance of Nic Wilson, my editor at OUP, who urged me to include lots of instructive examples in the book - it is the richer for it - and warmly thank her colleague James Cox for skilfully piloting me through the publication process.

Endorsements

'This is a timely and provocative analysis. It is crucially important that researchers, clinicians, industry, and regulators adapt effectively to the new era of stratified medicine. This book is a welcome contribution to the debate.'

Professor Jeremy Farrar, Executive Director, The Wellcome Trust, UK

'Richard Barker presents a provocative and thoughtful review of the health care ecosystem, demonstrating time and again that communication and collaboration across the stakeholder community are essential to realizing patient benefits from biomedical innovation.'

Dr Barbara Lopez Kunz, Global Chief Executive,
Drug Information Association, Switzerland

'Richard Barker, drawing on a quarter of a century of deep practical involvement in the topic, has written an invaluable synthesis of the factors that account for the disappointing rate of translation of science into affordable treatments to address unmet clinical need. Crucial will be forging of transparent productive relationships between academia, industry and an enlightened regulatory system. Essential reading for all those interested in making the medical innovation process faster, more productive and sustainable.'

Professor Sir John Tooke, co-Chair of the Centre for the Advancement
of Sustainable Medical Innovation, UK, former Vice-Provost of University
College London, and former President of the UK Academy of Medical Sciences

'Concisely frames the rich, dynamic environment for biologic/medical science with all the opportunities and challenges and offers orderly, thoughtful, practical approaches and solutions to facilitate and optimize medical innovation science. This book is suitable for both the knowledgeable scientific and the interested lay communities.'

Dr Michael Friedman, Emeritus Chief Executive, City of Hope Cancer Center,
Duarte, CA, USA and board member of life science companies

'This book provides an overview and in-depth analysis of medical science, drug development, etc., along with recommendations that are impressive. Amazing to say the least.'

Mike Casey, board member and former Chief Executive of US biotech companies and former senior executive at Johnson & Johnson

'This is an important and timely book. Richard Barker comprehensively reviews current progress, but perhaps more importantly points out encouraging paths of translation as we move forward. In doing so, he provides a comprehensive scientific and socio-cultural rationale and basis for closing the "collaboration gap", for bridging scientific advances, new product development and evaluation, and clinical care.'

Stephen P. Spielberg, MD, PhD, former Dean, Dartmouth Medical School, Hannover, NH, USA, former Deputy Commissioner, US FDA, USA, and Editor-in-Chief, Therapeutic Innovation and Regulatory Science

Contents

Contents

Abbreviations

ACE	angiotensin converting enzyme-inhibitors	GA4GH	Genomics Alliance for Global Health
ALS	amyotrophic lateral sclerosis	GCP	Good Clinical Practice
APC	Advanced Purchase Commitment	GDP	gross domestic product
		GGP	Good Genomic Practice
ARB	angiotensin receptor blockers	GIST	gastro-intestinal stromal tumour
ARM	Alliance for Regenerative Medicine	GLP	Good Laboratory Practice
		GPCR	G-protein coupled receptor
AZT	azidothymidine	GPS	global positioning system
BMI	body mass index	GWAS	genome-wide association studies
CASMI	Centre for the Advancement of Sustainable Medical Innovation	HAART	highly active antiretroviral therapy
CDC	Centers for Disease Control	HCV	hepatitis C
CFTR	cystic fibrosis trans-membrane conductance regulator	HPV	human papilloma virus
		HTA	Health Technology Assessment
CHMP	Committee for Medicinal Products for Human Use	HTLV	Human T-cell lymphotropic virus
CML	chronic myeloid leukemia	ICD	cardioverter-defibrillator
CNS	central nervous system	IMI	Innovative Medicines Initiative
COPD	chronic obstructive pulmonary disease	IPO	initial public offerings
		iPOP	integrated Personal Omics Profile
CT	computed tomography	ISPOR	International Society for Pharmacoeconomics and Outcomes Research
CTSCC	CASMI Translational Stem Cell Consortium)		
DC-MRI	Dynamic MRI	IT	information technology
DRG	Diagnostic Related Group system	MAPK	mitogen-activated protein kinase
EGFR	epidermal growth factor	MAPPs	Medicines Adaptive Pathways for Patients
EMA	European Medicine Agency		
ePS	endogenous pluripotent somatic	MDS	myelodysplastic syndrome
ERK	extracellular signalling-regulated kinase	MHRA	Medicines and Healthcare products Regulatory Agency
EUPATI	European Patients' Academy on Therapeutic Innovation	MMR	measles mumps rubella [vaccine]
		MRI	magnetic resonance imaging
FDA	Food and Drug Administration	NDA	new drug approvals
FIPCO	fully integrated pharmaceutical company	NIH	National Institutes of Health
		NRT	nucleoside reverse transcriptase
FNIH	Foundation for the National Institutes of Health	PCT	Patent Cooperation Treaty

PD	pharmacodynamic
PET	positron emission tomography
PK	pharmacokinetic
PROM	patient-reported outcomes
QOF	Quality Outcomes Framework
R&D	research and development
RA	Rheumatoid arthritis
RCT	randomized clinical trial
RWD	real-world data
RWE	real world evidence

SCG	Structural Genomics Consortium
SME	small and medium-sized enterprise
SSRI	selective serotonin re-uptake inhibitors
STED	stimulated emission depletion
TNF	tumour necrosis factor
VCP	valosin-containing protein
WES	whole exome sequencing
WGS	whole genome sequencing

Chapter 1

Introduction: The growing innovation gap in life sciences

At the heart of this book is a mystery. Our everyday experience is of accelerating innovation in many areas that touch our lives. In information technology (IT) and services, we live in an exponential age. But in the area of perhaps our greatest interest—our health—innovation seems stubbornly slow.

Before we dive into the mystery further, let us step back and ask some basic questions. What is true innovation?—and are we really living in an innovative society?

Are we living in an innovative 21st century society?

I'm writing a book on innovation on a 21st century product—a sleek laptop—sitting at a 20th century product, a desk designed by someone described as a 'prolific architect and innovator', Osvaldo Borsani. So perhaps we should start with a question: What makes a product innovative?

The MacBook Air, with its slim profile, large sharp screen, and well-designed keyboard, looks innovative, although we can clearly trace its evolution from the desktop computers that preceded it. It combines a great deal of technology in a small space and results from a rapid sequence of incremental improvements since the first personal computer, the breakthrough innovation. Is the desk innovative? Well, the gently curved shape is practical, so I can reach everything without moving my seat. It has a cunning way of locking the drawers. And a small bookshelf built into it—but at the back, where I can't reach from here. But in most respects, this desk is just a place to sit down and write, and its basic features—flat top, four legs, drawers to put things in—haven't changed for centuries. In that basic sense, this desk, attractive as it is, isn't really an innovation, other than aesthetically. Which, of course, is a benefit.

So what is innovation? Innovation, as I will use the term, is the ability to turn new science and technology into practical things—products and services—that are better, faster, or cheaper solutions to human need. The only potential innovations that really count are ones that are translated into beneficial impact for the ultimate customer.

So are we living in innovative times? I have heard commentators like Gary Kasparov, the chess champion, and Peter Thiel, the entrepreneur venture capitalist, say that the conservative nature of our institutions is too often stifling our innovative drive. But it is undeniable that, in many areas of human need, we have seen innovation deliver benefits better, faster, and often also cheaper. The smartphone sitting beside me is seen by most of us as a major breakthrough, even if it actually is mainly the bundling of past innovations—many of which are 20–30 years old—into one single, portable, ingenious device. And this smartphone is just the tip of a veritable iceberg of IT innovation, with wireless networks, cloud computing, apps, and GPS tracking all underpinning its success.

Across the board, innovation in IT is speeding ahead, constantly accelerating in its pace. The gap in time from the appearance of a new device or service and its widespread adoption is reducing with every new generation. The cost per new unit of capability is falling at a rate that is faster than exponential. Computer speed is a case in point: the number of calculations per second per $1000 of computer investment has risen from 1 in the 1950s to 10 000 000 000 today—10 to the power of 10 in my lifetime.

This has been powered by steady advances in a single basic technology, the microchip. Since it was first published in 1965, Moore's Law—the doubling of the density of transistors on a microchip every 2 years—has been followed faithfully by the semiconductor industry. And we have yet to see the real impact of optical or quantum computing! At the current course and speed, it is estimated that, by around 2045, computing will deliver (in 1 second for $1000) 10^{26} calculations, roughly the number performed by all the human brains on the planet.

Since its inception, the Internet has also grown exponentially. As the result of the subsequent invention of the World Wide Web opening up millions of sources of content to billions of users, it now reaches more than a third of the world's population (Figure 1.1).

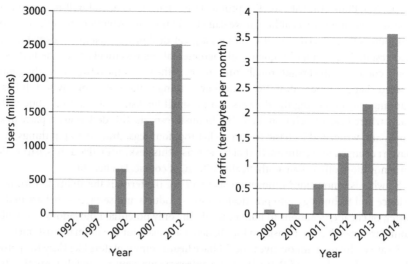

Fig. 1.1 Growth in Internet users and traffic.
Source: Data from Internet Live Stats, http://www.internetlivestats.com/, accessed 01 Oct. 2015; Cisco Visual Networking Index (VNI). 'Global IP Traffic Forecast 2009–2014', http://www.cisco.com/c/dam/en_us/about/ac78/docs/Cisco_VNI_Global_and_APAC_IP_Traffic_Forecast.pdf, accessed 01 Oct. 2015. Copyright © 2009 Cisco VNI.

This in turn has driven a truly astonishing growth in traffic, which has virtually doubled every year, to the mind-numbing level of 3.6 million terabytes per month (Figure 1.1). By the way, just 10 terabytes could hold the whole Library of Congress (itself holding more than 30 million books).

IT patents, too, are rising exponentially, promising yet further practical advances, from smartphones that advise and monitor us in real time, in all areas of our life, to artificial intelligence to make sense of all the data we are accumulating. The IT specialists IDC estimate a 50% increase year on year, so that we will be the lucky generators of 40 zettabytes of data by 2020, more than 50 times the amount we have today. This data will increasingly be unstructured data—including videos, images, voice records, blogs—in fact anything that doesn't come fitted within a structured database. These growing stores of data will in turn drive further innovations to mine and make sense of it all.

The speed of change in information businesses is breathtaking in relation to 20th century experience, when products and industries lasted for decades. Film-based photography originated in the 1820s, but digital photography replaced it in 10 years. In an ironic twist of fate, digital photography was first developed by Steven Sasson in Kodak, the company it went on to bankrupt. In the process, the technology has gone from delivering 0.01 of a mega-pixel for $10 000—via a device weighing 1.7 kg—to 10 mega-pixels for $10 and 0.014 kg! In short, 1000 times the resolution for 1000 less in cost and weight. A billion-fold improvement in performance.

The GPS receiver, that we are now all rely on (although we may have no idea of how it functions) began life in 1981, costing $119 900 and weighing 24 kg. Today, the same functionality is delivered for less than $5 in a device that is less than a centimetre across. As well as quantity, the quality of what is delivered to consumers advances also. Devices—from HD TV screens to embedded cameras in tablet computers—have become steadily more impressive in performance.

Companies involved in the digital revolution also grow (or disappear) more and more rapidly. Remember Netscape? 'Exponential' is a favourite word in Silicon Valley, and many of these companies and their technologies exhibit exponential growth, at least for a while. A few, of course, remain decade after decade, adapting as best they can to the changing environment. From the first successful mainframe computer in the 1960s, IBM developed Big Blue to defeat the world's leading chess player Gary Kasparov in 1997. Now its artificial intelligence successor, Watson, can mine sources on a vast variety of subjects and come up with findings that no human being has the brain capacity or time to uncover. It can also beat any human contestant at Jeopardy.

I recently heard a talk by Peter Diamantis (the founder of the X-Prize). Included in his long list of 'exponential technologies' were: sensor networks, robotics, 3D printing, nano-materials, and artificial intelligence. Also, interestingly, synthetic biology, of which more later. So, in anything touched by the digital revolution, we are seeing innovation at a rate never before seen. In that sense, 21st century society is certainly very innovative.

However, we see science-driven innovation in other, less obviously 'digital areas'. The energy sector, long dominated by decades-old oil extraction, coal mining, and turbine technology, has seen fracking, solar energy, and electric cars all arrive in earnest in the last 10 years. Electric vehicle sales are taking off, brands proliferating and costs falling: the US sales are up more than six-fold over the last 3 years.

The emergence of green energy as a competitive force has also been dramatic, with solar energy moving from a useful source for remote installations in tropical areas to a mass production technology, with field after field in temperate parts of Europe now covered with solar panels. As the efficiency of photovoltaic cells has improved from 2% in 1955 to 40% today, the cost dropped from $1785 to $1.25 per watt, following the solar energy version of Moore's Law—Swanson's Law.

Moving from energy production to its consumption, after languishing for decades, the electric car is now entering a phase of exponential growth, propelled by environmental concerns and efficiency savings.

Even the highly conservative area of food has seen the application of chemistry by chefs such as Heston Blumenthal to create dishes never seen or eaten before. And many restaurant menus throughout the world contain creative fusions of cuisines old and new. On a more industrial scale, genetically engineered crops, controversial as they are in some parts of the world, are delivering more affordable food to millions.

So information processing, communications, food, transportation, and energy— some of the areas of greatest human need—are certainly feeling the touch of innovation. As the underlying technology advances, it feeds through to a rapid development of new products and the delivery of greater consumer benefit, often at an accelerating pace. And all of this translates into rapidly falling prices per unit of performance.

The net result of exponential innovation is often major disruption of industries: Uber, the world's largest taxi company, owns no vehicles; Facebook, the world's most popular media owner, creates no content; Alibaba, the most valuable retailer, has no inventory; Airbnb, the world's largest accommodation provider, owns no real estate. Of course, all of these new businesses are internet-enabled.

Exponential progress in innovation comes when three major factors combine. Governments lay the foundation by targeted investments in both basic science and early development, just as the US government did in the early days of the internet. Then start-ups and subsequently major corporations pour investment into R&D to develop sequential generations of products. And then consumers adopt and upgrade these, seeing ever-increasing benefit at lower or steady prices. Peer prompting encourages them to replace good products with even better ones.

Of course, there remain many conservative industries (when did you last see a major change in French bakeries, for example?), but for many areas of life rapid change seems now to be the norm.

Is this happening in life sciences?

When we turn to the life sciences and their impact on medicine, the pattern is not just different but startlingly so. While customers for IT innovation, for example, are enjoying faster, better, *and* cheaper products simultaneously, in biomedicine this is far from the case. Too often we are seeing products that make a fairly marginal impact, take many years to arrive, and end up costing more.

The delay between a basic bioscience breakthrough and its widespread impact on human health stays stubbornly at 15–20 years. The cost of new drugs and devices is spiralling upwards, not downwards. Moreover, with some exceptions, the impact on human health they bring is more incremental than revolutionary. Most products alleviate symptoms or extend life by a few months, rather than reverse disease or cure it. So customers for life science innovation—patients and payers alike—are not seeing products that arrive faster nor cheaper, and they are often not so much better either.

This is certainly not for want of trying, and investing. The pharmaceutical industry, for example, invests in R&D at twice the rate (in terms of percentage of sales) of the IT industry. Many companies spend 15–20% of their revenues on R&D, year after year. Medical device R&D continues to rise rapidly, now standing at $23bn annually.

Despite the heavy expenditure, the number of pharmaceutical products reaching the market has remained fairly stable, at 25–35 a year (as measured by Food and Drug Administration [FDA] approvals). So the productivity measure of products per unit of investment has fallen dramatically. My Centre for the Advancement of Sustainable Medical Innovation (CASMI) colleague Jack Scannell coined the phrase 'Eroom's Law' to describe this phenomenon: that the products delivered per unit of investment has demonstrated the *reverse* of Moore's law—exponential decline[1]. This is not just over the last 20 years, but for more than 50, and remains even when we adjust for the delay between investment and products appearing (Figure 1.2).

The result of this dramatic loss of productivity is that the investment required per new drug has reached several billions of dollars. Matthew Herper of Forbes recently calculated the cost for individual major companies, which varied from $3.3bn to $13.4bn![2]

Academic studies confirm the trend: the average cost, including the cost of money over time, has risen exponentially. The most recent (December 2014) figure for this from the group that has tracked it over many years, the Tufts Center for the Study of Drug Development, is $2.6bn[3]. The difference between this figure and the simple ratio per major company is consistent with the fact that smaller companies manage to spend less per new drug than most of the larger companies. So the contrast in innovation productivity with other industry sectors couldn't be sharper!

This is one of the major factors that has led to steady price rises for new drugs. Payers are proving increasingly reluctant to fund products they see as having modest value for high cost, even though they understand that most of the $2.6bn is the cost of failures: products that never make it to the market. We will analyse the productivity problem in much more depth in Chapter 3.

Now, what makes these trends so interesting is that they are in startling contrast to the rapid progress being made in the underlying science of human biology, the subject of Chapter 2. Genomics, proteomics, structural analysis, systems biology, cell biology, and medical imaging—the list of new tools and discoveries goes on. In all these basic biological science areas, we have seen huge advances over the last 20 years. These are reflected in the almost exponential rise in publications in life science journals, and the steady increase in patents filed. So, while scientific inputs have multiplied, outputs—in terms of products and patient benefit—have not.

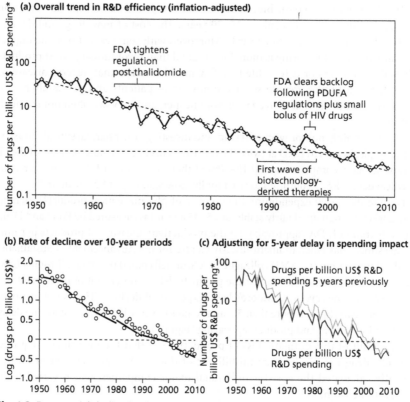

Fig. 1.2 Exponential decline in pharmaceutical products produced per unit of investment. Reproduced with permission from Scannell JW, Blanckley A, Boldon H, and Warrington B. Diagnosing the decline in pharmaceutical R&D efficiency. *Nature Reviews Drug Discovery*, Volume 11, pp. 191–200, Copyright © 2012 Macmillan Publishers Ltd.

The mystery deepens. The period in which many of today's most successful drugs were conceived—one that some call the 'golden age' of drug discovery—was not blessed with most of what now know of human biology, and most of the tools we can now use to probe it. We knew precious little about the complex gene control mechanisms, how proteins interact together in cellular networks, and how cells communicate. Sometimes all we knew was that a particular protein was somehow involved in disease, that the drug could inhibit or stimulate its activity, and that early experiments in animal models of human disease had seen a good response. In some cases, we simply found serendipitously that a drug that was being investigated for disease A actually had strong activity in disease B.

Twenty years on, we can sequence genes rapidly, profile their expression, determine protein structures, map protein networks, create huge libraries of promising compounds and probe their effects in high throughput experiments. So why has our innovative output not also increased by leaps and bounds?

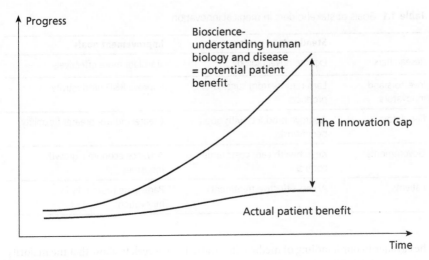

Fig. 1.3 The innovation gap.

In contrast, we see a widening 'innovation gap' (Figure 1.3).

This is the central question I will tackle in this book: what is it about the medical sciences and their application to achieve patient benefit that has defied the 'better, faster, cheaper' pattern of innovation seen in so many other sectors—and can anything be done about it?

We will look at the quality of basic science, efforts to translate it into potential products, and the process of development via clinical trials. We will examine the tests applied by regulators and payers to understand if they are reasonable or could be made more receptive to high value innovation. We will also explore the various barriers that even approved innovations can face in their uptake by conservative health systems, cautious clinicians, and confused patients. We will look not only at the origination of products but their spread through the system. This is remarkably slow, requiring 10–15 years in many cases, unless a significant incentive or driving programme is in place. The English National Health Service (NHS) itself quotes 17 years as a typical time![4]

So, despite huge advances in the underlying science, product origination is increasingly costly, R&D productivity has steadily fallen and those products that do emerge take many years to be adopted. A depressing pattern, in contrast to so many other science-driven sectors.

Some might find this comparison very simplistic, and say: of course medicine is different—it's more complex, deals with life and death issues, and should properly progress slowly and cautiously. But these people are unlikely to be patients with severe, life-limiting conditions.

The basic question is this: does our slow and costly medical innovation process simply reflect basic knowledge gaps or intractable aspects of human biology, or are they a reflection of human choice—how research is organized, products developed, innovation rewarded, risks assessed, regulations drafted, and users engaged? While there are indeed some huge gaps in our basic knowledge of biology, and we can never afford to

Table 1.1 Goals of stakeholders in medical innovation

	Steady-state goals	**Improvement goals**
Researchers	Explore novel biology	Translate more effectively
Investors and innovators	Earn returns from successful products	Improve R&D productivity
Regulators	Approve products with good benefit/risk	Create and use greater flexibility
Governments	Keep health care costs under control	Advance economic growth industries
Patients	Access effective treatments	Participate more fully in innovation

be cavalier in our handling of medical innovation, I will seek to show that the majority of factors that inhibit the medical innovation process *are* subject to human choice. So, if we understand our choices and their implications we can change them. That is the good news.

One complexity is undeniable. Medical innovation involves an unusually wide range of stakeholders: academic researchers, biopharmaceutical companies, device and diagnostics manufacturers, regulators, health technology assessment (HTA) agencies, politicians, clinicians, payers (public and private), and last, but of course far from least, patients and those who care for them. They all have different goals, both in terms of their basic needs and the improvements they would like to see (Table 1.1). An effective innovation process must satisfy the needs and concerns of all. Throughout this book I will seek to reflect these different perspectives, both in the specific case examples I discuss and in terms of general principles for the future.

What approach am I taking?

Firstly, the problem has many causes and therefore needs several different and complementary solutions. Most problems are a function of the overall system, not specific players: we need to avoid the 'blame game'. Some complain about academics uninterested in turning their discoveries into useful therapies. Some lay responsibility for disappointing productivity at the door of conservative regulators. Some level their criticism at avaricious pharmaceutical companies unwilling to invest in difficult disease areas. Others blame short-sighted HTA agencies unwilling to recognize the future value of innovation as they attempt to control prices. Still others point to the conservatism of clinicians.

In contrast, I will assume all the players are seeking to play their roles according to their mission and strengths and that we should look to change how they all interact in what is increasingly called the 'innovation ecosystem'. For although all have a critical role to play—academics, companies, regulators, reimbursement agencies, doctors, and payers—they do not couple together to form an effective system. They leave what I describe as major 'gaps in translation'.

In the course of the book, I will also challenge a number of popular myths about the innovation process. No one person will hold all these beliefs, but collectively they impede our progress:

- Academic science produces all the major innovations; then companies patent them and reap all the rewards.
- The products that are in the pipeline are mainly 'me-toos' of marginal value.
- Life science companies are sworn competitors and will never cooperate.
- Patents are the basic problem: if we abolished them we would get more innovation, more cheaply.
- Regulators are much more conservative than the companies with whom they deal.
- The medical innovation business, because of its complexity, is best left to the professionals: patients can contribute little.
- Doctors always know best and are able to keep up with their fields by reading medical journals.
- Patients often do not comply with their medications because they simply forget.

The subject of this book—creating a 'sustainable ecosystem'—sounds academic and theoretical. But it's far from that, for at least two pressing reasons. First, many of the people that fund the life sciences—from governments to company investors—are seeking greater returns, in terms of practical innovations and valuable health advance. And, despite the exciting scientific headlines, it is the delivery of practical outcomes that determines long-term investor confidence. The whole innovation enterprise is in jeopardy if disappointed investors, public and private, cut back on the research that fuels it. One key measure of confidence is the amount the global pharmaceutical industry spends annually on R&D: this rose steadily in the 1990s and early 2000s, reaching $136bn by 2011. It actually fell slightly in 2012, and has now resumed growth, but at a much reduced level.

The most important reason to close the innovation gap waits for treatment in doctors' surgeries, outpatient departments, and nursing homes across the world: patients with unmet need. There are literally thousands (at least 7000) of untreated rare diseases, scores of still-fatal cancers, the steady neurodegeneration of millions of Alzheimer's and Parkinson's sufferers, and the human tragedy of premature deaths from such unconquered diseases as muscular dystrophy and amyotrophic lateral sclerosis (ALS). Most treatments for chronic conditions such as chronic obstructive pulmonary disease (COPD), heart failure, osteoarthritis, and diabetes we have today deal with the symptoms, not the disease itself, let alone its prevention. The work of medical innovation is vital and far from over.

In tackling the issue of the innovation gap, the book's chapters flow as follows. Hopefully by this point (Chapter 1) you are persuaded there is a problem to tackle. Chapter 2 reviews the dramatic scientific advances that have occurred in the last 20 years and should be forming the basis of medical innovations today and tomorrow. Chapter 3 will analyse the 'gaps in translation' that are preventing the full benefit being received of this revolution in bioscience. Chapter 4 will scan a host of examples of both past success and failure to seek to distil lessons on what can and should change about the innovation process.

Chapter 5 is the core of the book's conclusions and recommendations on how we should reform and refashion medical innovation. It lays the scientific foundation for seven steps to sustainability, with their major unifying theme of precision medicine.

However, the environment needs to be right for the changes to occur and be embraced by wider society: Chapter 6 will review aspects of this environment and point to supportive changes needed in both policy and practice. Part of the argument for change, both in the process and the environment will be a sense of the prize: Chapter 7 will describe the impact of both the enabling technologies and disease programmes we could see in the next 20 years if we get the changes right.

Finally, in Chapter 8 we will step back and survey the overall 'innovation ecosystem' and the revolutionary changes to roles and relationships that the changes will imply.

Although I've taken pains to emphasize the challenges we are facing in turning medical discoveries into medical advance, I want to strike a very positive note at the outset. There are promising signs of a turnaround beginning. The last 2–3 years have seen a resurgence in new products approved by the FDA: most of these have been successful because they are better targeted. There is much debate on how sustainable this recovery is, but it has restored investor confidence, at least in the small and medium-sized enterprise (SME) sector. There have been more than 120 initial public offerings (IPOs) of biotech companies, and more established companies with products in Phase 3 trials have been able to sell themselves to the majors at very high prices. However, investor sentiment in biotech is notoriously fickle, and much depends on the success of the pipeline. The only way this recovery in investor confidence can be maintained is for promising products to make it through to approval, reimbursement, and use. The challenges at each of these stages have not changed.

I also see many examples of the various players in this drama beginning to take new stances and play new roles. There is a rapid increase in what is called 'pre-competitive' collaboration between companies. They are collaborating on areas that they realize are impossible—or just very inefficient—for them to pursue alone.

Regulators in many parts of the world are showing a new level of flexibility and responsiveness in dealing with novel types of innovation or therapies in areas of high unmet need. HTA agencies are asking searching questions about their methodology. Payers are prepared to consider new types of commercial arrangements that reward innovators for impact, not just supply of product.

Perhaps most importantly, patient organizations are leaving their comfort zone of highlighting need and lobbying for treatment to actually design new therapies and participate actively in the innovation process. It is always the patient with most at stake, and industry, the regulators, the clinicians, and the payers ignore them at their peril—and the politicians know their votes count!

So, with early promising signs of change in attitudes as well as practice in many quarters, it is a good time to review what we need to do to create sustainable innovation.

Some final comments before we start

This is not a 'how to' book on the basic principles of the discovery and development of medical technologies. There are some useful accounts of this subject[5,6,7]. There is also an excellent and provocative recent book on the specific challenges of drug research by Tomas Bartfai, a lifelong drug discoverer, that I commend to readers[8].

I will analyse many case examples of innovation, but my goal is to go beyond anecdotes and wherever possible to take a robust scientific approach to the problem. My colleague Rob Horne of UCL coined the useful term 'innovation science' for the study of the innovation process itself, the factors that affect it, and the stakeholders for whom they are important. My intent in the pages that follow is to lay some of the foundations of a new discipline of 'medical innovation science' for researchers and innovators in both public and commercial R&D organizations. Understanding this science will also have value for policy-makers and for the leaders of health systems. They desperately need faster, better, and cheaper innovation, if they themselves are to be sustainable. They will not be able to cost-cut their way to sustainable healthcare, as I have pointed out elsewhere[9].

Readers will notice there is proportionately more diagnosis and prescription in the book for the translation gap in pharmaceuticals than for other parts of the life science sector. This reflects both the level of debate and the seriousness of the issues. But wherever possible I will compare and contrast the situation in pharmaceuticals with that in medical devices, in vitro and in vivo diagnostics, and in digital health, or devote sections to these vital elements of the sector.

Finally, I will summarize at the end of each chapter its key conclusions, and how they lead into the next chapter. This is intended to help readers who may, for example, be familiar with the science (they may wish to skip Chapter 2), or with the barriers to translation (Chapter 3), and wish to move straight to the case studies or my proposed solutions.

Summary of Chapter 1

In a world caught up in ever more rapid technological advance, biomedical innovation remains stubbornly slow and unproductive, as measured by output (benefit to patients) over input (investment). This is despite very rapid progress in the underlying science, and it puts at risk the very enterprise of life sciences, in which society and investors place so much faith. We need a scientific study of the gaps in translation, and radical thinking to bridge them.

In Chapter 2, we review the breathtaking advances in basic bioscience and related technologies over the last 20 years—advances that have the potential to lead to major innovations, if we can reform the innovation process.

References

1. **Scannell JW, Blanckley A, Boldon H, Warrington B.** 2012. Diagnosing the decline in pharmaceutical R&D efficiency. *Nature Reviews Drug Discovery*, 11: 191–200.

2. Matthew Herper, *Forbes Magazine*, http://www.forbes.com/sites/matthewherper/2013/08/11/the-cost-of-inventing-a-new-drug-98-companies-ranked/

3. Tufts Center for the Study of Drug Development report, dated 18 November 2014, http://csdd.tufts.edu/news/complete_story/pr_tufts_csdd_2014_cost_study

4. NHS England, 'Health and high quality care for all, now and for future generations: Frequently asked questions', http://www.england.nhs.uk/ourwork/innovation/nia/faqs/, accessed 01 Oct. 2015.

5. Burns LR. 2012. *The Business of Healthcare Innovation*, Second Edition, Cambridge: Cambridge University Press.

6. Rang HP. 2006. *Drug Discovery and Development*, Oxford: London: Elsevier.

7. Barker R, Darnbrough M. 2007. The role of the pharmaceutical industry. In: Kennewell PD, Moos WH, Kubinyi H, et al. (eds), *Comprehensive Medicinal Chemistry II*, vol. 1, pp. 527–52.

8. Bartfai T, Lees GV. 2013. *The Future of Drug Discovery: Who Decides Which Diseases to Treat?* London: Academic Press.

9. Barker R. 2011. *2030—The Future of Medicine: Avoiding a Medical Meltdown*, Oxford: Oxford University Press.

Chapter 2

The accelerating pace of biomedical advance

The problem in life sciences is not slow scientific progress. In this chapter, we highlight and celebrate the most tremendous advance in biological understanding in human history, one that has occurred over the last 20–30 years. As we apply ever more precise techniques to probe biology at the cellular and molecular level, we are discovering both the massive complexity of our bodies and their diseases, but also fundamental mechanisms we can use to create powerful interventions.

In this chapter, we will devote sections to charting this progress in genomics, epigenomics, protein structure and function, and the systems biology that ties these together. Major advances have occurred in tools such as biomarker discovery, imaging, the laboratory development of stem cells, and even the 'creation' of cells via synthetic biology. And the process of synthesizing molecules that can specifically target disease-associated proteins has also been revolutionized and indeed industrialized.

The development of these fields has been both qualitative—new avenues of research opening up each year—and quantitative. I will illustrate the latter with a series of charts that track (via Google Scholar) the number of papers in each field over the last 20–25 years, using key words on the topic. These are, of course, a crude measure, but they serve to show the rate of growth over recent years.

The sequencing of the human genome just 15 years ago was only one major milestone in the process of unravelling the secrets of life. The control of gene expression by epigenetics, the structure and function of proteins and the pathways that connect them, and our understanding of the structure and behaviour of cells have all seen groundbreaking discoveries celebrated by a sequence of Nobel prizes.

Many of the 21st century Nobel Prizes in Medicine have obvious potential relevance to human disease: cell signal transduction (Carlsson, Greegard, Kandel, 2000); the understanding of the cell cycle (Hunt, Nurse, 2001); genetic regulation (Brenner, Horvitz, Sulston, 2002); magnetic resonance imaging (MRI) (Lauterbur, Mansfield, 2003); discovery of the role of *Helicobacter pylori* in ulcers (Marshall, Warren, 2005); gene silencing by RNA (Fire, Mello, 2006); embryonic stem cells (Capecchi, Evans, Smithies, 2007); discovery of the viruses HPV and HIV (Hausen, Sinoussi, Montagnier, 2008); the role of telomeres in DNA (Blackburn, Greider, Szostak, 2009); in vitro fertilization (IVF) (Edwards, 2010); innate and adaptive immunity (Beutler, Hoffmann, Steinman, 2011); and development of pluripotent stem cells (Gurdon, Yamanaka, 2012).

Most of these discoveries are leads or tools for product development in areas as diverse as cancer and infectious disease. There is usually a gap of several years between the research and the award of the prize, and so we would expect now to be seeing the fruits of this work in practical treatments. Certainly we do in some: ulcer therapy, IVF, and the application of MRI.

One of the most notable prizes of the 20th century was, of course, awarded to Crick, Watson, and Wilkin for describing the structure of DNA, and the genome has been the focus of huge research interest ever since.

Understanding our genome

From the first human genome sequence published in 2000, DNA sequencing efficiency, in terms of cost and speed, has increased at a breath-taking pace. Over the last three decades, the rate of DNA sequencing has accelerated by around 2 billion times[1,2,3]. Correlating genomic variation with human disease, via genome-wide association studies (GWAS) has proceeded apace. The number of such associations increased from a single such finding in 2005 to many hundreds by 2013[4].

With the development of rapid sequencing methods, the cost of whole genome sequencing (WGS) has also dropped dramatically, from the $3bn needed for the first full sequence in 2000 to approximately $1000 today. Over the last few years the cost has dropped much faster than Moore's Law for semiconductor performance (Figure 2.1).

So we are now far exceeding IT in the pace of advance in the most basic of the bio-technologies that should power our progress. On this basis, we have developed rapid and affordable 'exome' sequencing, focused on those sections that code for proteins.

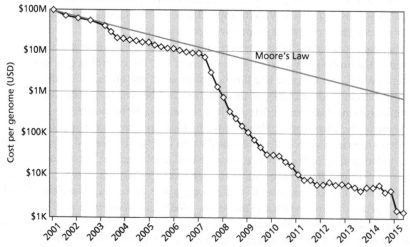

Fig. 2.1 Rapidly falling cost of gene sequencing. Estimated genome sequencing cost (2001 to 2015).

Reproduced by kind permission of National Human Genome Research Institute, http://www.genome.gov/sequencingcosts/, accessed 01 Sep. 2015.

We have also begun to explore the genome beyond the 2% of it that codes for proteins, much of it originally termed 'junk DNA', thought essentially to be along for the ride in evolutionary terms. We begin to understand the function of the large sections of DNA that code for RNAs, vital for control of transcription. The amount of DNA thought to be non-functional is falling rapidly.

WGS is now seen as affordable, and is ready to move from being a research tool to being applied in clinical situations, or perhaps as something performed on every new-born. We have identified over 50 million single-nucleotide polymorphisms, the single base differences between human individuals, and 3000 disease-associated genetic mutations.

We have not restricted ourselves to the human genome: we have now determined over a hundred vertebrate genome sequences, about five hundred non-vertebrate species, and nine thousand prokaryotic genomes. This enables us to spot homologies (common regions) and track evolution more scientifically than ever before.

The advances in time and cost have unleashed a huge expansion of sequencing activities, from determining the sequence of individual clones within a tumour to a recent effort to 'sequence' the New York subway. (The latter, incidentally, looked for fragments of DNA in swabbed metro handles and other carriers of human and microorganism DNA in the subway system, and found huge amounts of unidentifiable genetic material, and fragments of some pathogenic organisms!)

However, much of the more serious genomics work has focused on cancer, where mutations are clearly central to the development of the disease and multiply as the cancer evolves. Distinguishing between 'driver' mutations that underlie the development of the cancer and those that simply result from loss of the normal repair mechanisms has become a challenge. We have also found many genomic relationships between different cancers that start in different organs but have the same underlying mechanisms.

The growth of such studies has been explosive, as genomic tool costs have plummeted and the availability of convenient benchtop sequencers has mushroomed (Figure 2.2).

The UK government's programme to sequence 100 000 genomes from its citizens, with cancer a major focus, for a comparatively modest £200 million budget, is clear evidence of just how far we have come in a very short space of time. There is little doubt that this exponential development of understanding our genetic makeup—and how it goes awry in cancer—will continue, as costs continue to decline.

As mentioned above, part of the mystery of 'junk DNA' is explained by its coding not for proteins but for RNA with a wide variety of roles in the cell. Long non-coding RNAs form a key part of the cellular control mechanism and around 100 of these are found at very high levels in very aggressive cancers (12). Early studies of cancer biology emphasized the discovery of proteins like p53 expressed at a high level in tumours; we are now homing in on some of the non-protein factors at play.

In fact, our intense focus on the genome's role in coding for proteins has probably been unbalanced. The enzymatic properties of RNA transcribed from the genome are vital for the protein manufacturing mechanism. RNAs are also active as catalysts, promoters, and inhibitors of many of the other activities of the cell.

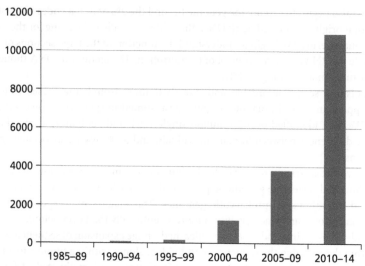

Fig. 2.2 Scientific papers on cancer genomics (up to 2014).
Source: Data from Google Scholar, [search term] 'cancer genomics', https://scholar.google.co.uk/, search date 31 Aug. 2015.

To identify which genes are being expressed at any one time and in any one type of cell we have the techniques of 'transcriptomics'. Specific RNA sequences are identified and quantified using gene chips, in a way that gives us a 'fingerprint' of a cell. This was the first approach to cancer genomics before exome or whole genome sequencing became available and affordable, and it still gives insights in areas such as cancer.

Overall, we have seen a successive of cumulative insights into the genome, its sequence, structure, and expression—and the pace seems likely to quicken further (Figure 2.3).

Methods and convenience of sequencing are at the heart of the genomics revolution. As the speed, cost, and convenience of alternative means of querying the genome evolve, we will see ongoing changes in their relative popularity. What is undeniable, however, is that we are in a period of high growth in all of them (Figure 2.4).

There are many mysteries of the functionality of the genome still to be solved. We will continue to apply the tools of genetic modification, evolutionary study, and, increasingly, biochemical activity to map the function of specific genes. These three complementary tools for probing the genome will finally resolve the continuing debate on the total proportion of the genome that is truly functional. But that proportion rises year on year. The international ENCODE project tracks progress regularly[5].

Cancer and rare inherited disorders are clearly leading the way in the genomic interpretation of disease. The genomics of complex, chronic diseases like diabetes will probably lag behind by several years, especially as non-genomic factors associated with human behaviour seem to be more significant in determining disease risk and development. Genotyping studies to date on the impact of 18 different genomic variants on diabetes risk have shown only weak predictive power, comparable to those based only

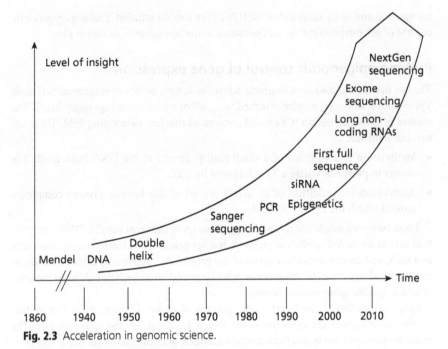

Fig. 2.3 Acceleration in genomic science.

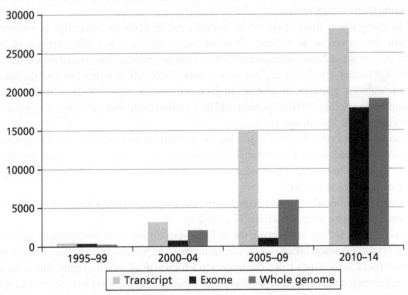

Fig. 2.4 Growth in genomic analysis technologies (up to 2014).
Source: Data from Google Scholar, [search term] 'transcriptomics'; 'exome sequencing';
'whole genome sequencing', https://scholar.google.co.uk/, search date 31 Aug. 2015.

on age, sex, and body mass index (BMI)[6]. In the section entitled 'Probing epigenomic control of gene expression' we will see some of the 'epi-genetic' factors at play.

Probing epigenomic control of gene expression

The last decade has also seen dramatic advances in epigenetics or epigenomics (Greek 'epi' = above, on, over, nearby, attached to … all of which meanings apply here). The control of gene expression is a central concern of this fast-developing field. There are two mechanisms:

- Methylation (i.e. addition of a small methyl group) of the DNA helix itself: this occurs in positions where a 'C' is followed by a 'G'
- Acetylation (i.e. addition of an acetyl group) of the histone protein complexes around which the DNA is wound.

These two very subtle chemical changes have quite different results: DNA methylation acts as an 'on/off' switch of genomic transcription, while histone acetylation acts as a finer, and more reversible, control of the process. So, apparently minor changes in the chemical structures of the genome drive large and often heritable shifts in the level at which specific genes are expressed.

Epigenomics lies at the heart of the process of tissue differentiation that distinguishes which genes are expressed—and which are silent—in the cells that form the liver, the skin, and the brain, from initially identical genomic material. But it also controls how—once in place in the tissues—the cells respond to everything from environmental stress to diet change.

In many cases, these epigenomic controls, put in place by behaviour or environment, are passed on to subsequent generations. Thus diet and other aspects of the environment can have consequences beyond the generation that experiences them.

Epigenomics also helps explain how genetically identical twins can and do suffer from different patterns of disease. In fact, monozygous (identical) twins leave the womb with slightly different patterns of DNA methylation, and the differences become greater as their lives progress.

So we have moved beyond regarding the genome as determining everything about us. We now see it as the raw material of life that gets modified, directed, and controlled by what happens to us and how we respond to it. Our own personal biological management challenge, if you will.

While the general concept of epigenomics or epigenetics has a long history, the last 5 years has seen a surge of activity (Figure 2.5).

As a result, we are uncovering the link between epigenomic mechanisms and several human diseases. Like genomics, epigenomics has particular importance in cancer, given that cancer arises from the uncontrolled growth of mutated cells. The interaction between epigenetics and mechanisms of tumour suppression has received a lot of attention recently. Hypermethylation of one of the genes involved in tumour suppression switches off its activity, and so helps drive renal carcinoma[7]. Inhibiting enzymes that methylate DNA shows promise as a route to drug discovery. 5-azacytidine, a potent inhibitor of such an enzyme is already an established drug in the treatment of myelodysplastic syndrome (MDS).

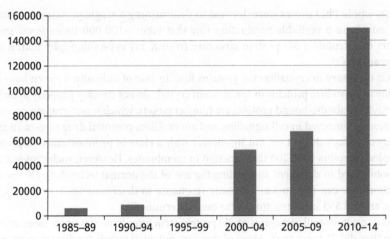

Fig. 2.5 Growth in epigenetics (up to 2014).
Source: Data from Google Scholar, [search term] 'epigenetics', https://scholar.google.
co.uk/, search date 31 Aug. 2015.

Some rare diseases have epigenomic causes. A striking example was one of the supposed cases of autism 'caused' by the MMR vaccination. The child, in fact, suffered from Rett syndrome, in which there is a mutation in the gene for the enzyme that reads methylated DNA.

Epigenetic mechanisms have now been discovered in a wide range of other disease states, including diabetic retinopathy, trauma-induced anxiety, cardiac dysfunction, drug addiction, and schizophrenia. It is even thought that epigenetics is responsible for much of the evolutionary change through time in species, challenging the previous model based solely on random genetic mutation[8].

So epigenomics controls genomic expression of proteins, the primary target of drug discovery. Equally rapid progress has occurred in protein biology.

Proteins—mapping their structure and function

We now know there are about 20 000 genes that code for proteins in our genomes. We have known about the molecular make-up of proteins—the strings of amino acids that determine their function—for more than 50 years. Probing their three-dimensional structures, the way they fold to create 'active sites' for their catalytic roles, has led to a true understanding of their function, and how they can be modified—blocked or unblocked—by drugs.

In parallel, we have developed the technology, based on immunoassay and mass spectrometry, to assess protein levels in various types of cells (the 'proteome') and the way they interact together in pathways, catalysing or inhibiting each other's function. As a result we have a fast-growing understanding of their role in health and disease.

In particular, there has been dramatic progress in *protein structure determination*. The first structure (of insulin) was determined in 1969 by Dorothy Hodgkin after months of painstaking analysis of X-ray crystallographic data. In subsequent years, whole PhD theses were devoted to determining a single protein's structure. We now have a veritable production line that takes ~100 000 times fewer man-hours to calculate a 3D protein structure from X-ray crystallography than it did 50 years ago[9,10].

It is necessary to crystallize the proteins first. In case of difficulty, we even have the ability to crystallize proteins in space, when crystals do not develop properly in earth's gravity! Membrane-bound proteins are tougher targets, which is unfortunate, as many of these are involved in cell signalling and are excellent potential drug targets. In fact, many of today's drugs turn out to interact with a class of proteins called G-protein coupled receptors (GPCRs) that function in membranes. However, such proteins can be solubilized in detergent and, failing the use of the normal technology (X-ray diffraction), we can use nuclear magnetic resonance to determine their structures. To date, around 550 such structures have been determined[11].

As a result, overall, databases of 3D protein structures have 300 times more entries than they did 25 years ago[12]. Many of these are potential targets for drug discovery.

It makes sense for protein structure determination to be conducted in 'pre-competitive space': in other words, making the results available, free of any patents, to any company or other drug discovery unit that believes they can design a molecule that fits the active site. Such sites have very specific requirements, in terms of shape, size, charge density, and hydrophilic (water-seeking) versus hydrophobic nature, but in practice many different 'drug' structures can fulfill the same requirements.

There are estimated to be 10^{62} possible permutations of carbon, hydrogen, nitrogen, and oxygen (the normal elements in a drug) to make drug-like molecules. This is more than the number of particles in the universe, so the possibilities are truly endless! Therefore starting with the same protein structure, research groups can devise widely divergent drug design strategies.

This philosophy is the basis of initiatives such as the Structural Genomics Consortium (SCG), a partnership of eight companies (Abbvie, Bayer, Boehringer Ingelheim, Janssen, Merck, Novartis, Pfizer, and Takeda), the Wellcome Trust, and a number of Canadian foundations. The SGC is an international organization, based both in Oxford and Toronto. They alone have been able to determine structures at a dramatic rate and currently have 1755 proteins in their 'structure gallery'[13]. They focus on proteins that their partners believe could be of potential interest for the development of new drugs, but have not attracted sufficient attention to date from other structure determination efforts.

As a result of all these efforts, we are now able to map a whole class of proteins that we know are central to the process of signalling within and between cells—the kinases. There are around 500 such proteins in human biology and—although a tiny percentage of the total—they are thought to interact with around a third of other proteins and so change the behaviour of most of the cell's pathways. We have already found the structures of nearly 200 of them[14] (Figure 2.6).

From Figure 2.6 it is clear that, while industry and individual academic groups continue to be active, the consortium approach is overtaking both in kinase structure productivity.

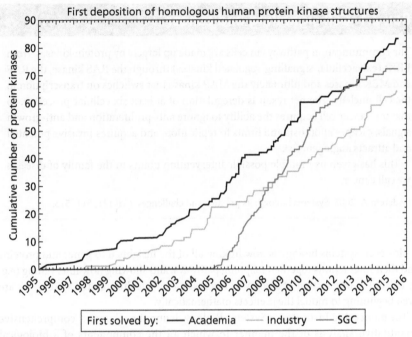

Fig. 2.6 Growth in kinase structures solved.
Reproduced with permission from SGC, 'Progress in protein kinase structural biology',
http://www.thesgc.org/scientists/resources/kinases, accessed 01 Sep. 2015, Copyright ©
2016 SGC.

We have learnt how proteins span cell membranes and hence transmit signals from outside the cell. The signalling pathways triggered by these external messengers enable the cell to adapt to its environment and 'read' what the cells around it are doing. We now have a particularly detailed map of the signalling pathways involved in most cancers (see Box 2.1). As a 2007 review put it: 'cancer can be perceived as a disease of communication between and within cells'[15].

The full understanding of the proteome—all the proteins, their function, structure, and interactions—is still a long way off. But it should be clear we have made enormous progress in this arena in the last 20 years, with a particular focus on those proteins we know, from gene or other studies, are involved in disease mechanisms.

Systems biology—the wiring diagram of life

With 20 000 genes, controlled by multiple epigenetic and other signals, coding for perhaps two to three times as many proteins, each of these interacting with several others in pathways, we obviously have a complex systems problem that will keep scientists busy for several generations. This is what the mathematicians would call a 'complex adaptive system' whose analysis—static and dynamic—will challenge researchers to probe and describe.

Box 2.1 Cell signalling and cancer

The communication pathways in cells are made up largely by protein kinases, from ERKs (extracellular signalling-regulated kinases) through the RAS kinase, the RAF and MEK kinases and ultimately the MAP kinase that switches on transcription in the cell nucleus. The net result is deregulation of at least six cellular processes, so that the cancer cell acquires the ability to ignore anti-proliferation and anti-growth signals, escapes apoptosis and limits to replication, and acquires invasive potential and attracts angiogenesis[a].

This has given us multiple possible intervention points in the family of diseases we call cancer.

[a] Aderem A. 2005. Systems biology: its practice and challenges. *Cell* 121, 511–513.

However, systems biology is now linking all of the insights into genes and proteins and their interactions together, mapping pathways, and showing how intervening (e.g. by a drug) in one section of a pathway has cascade effects through the system. We are even beginning to model these effects mathematically.

The pioneering Institute for Systems Biology defines its focus as a 'comprehensive quantitative analysis of the manner in which all the components of a biological system interact functionally over time … executed by an interdisciplinary team of investigators that is also capable of developing required technologies and computational tools'[15]. The new tools of genomics, transcriptomics, and proteomics give us the ability to quantify the various components in the system, to map how they relate to each other and how the network evolves over time. This in turn enables us to build computer models of the networks—and ultimately of the overall cell. When we have the latter, we will have the response to the National Science Foundation's 21st Century Biology challenge set in 2002—to build a computer model of the whole cell.

It should not surprise us that the challenge of modelling the behaviour and interactions of a network of perhaps 50 000 proteins and 350 metabolites will be a decades-long endeavour. However, fast progress is occurring in cancer systems biology, mapping all the mutations found in cancer and showing how they drive changes in the cell's signalling pathways. The Cancer Genome Atlas, being created by seven data analysis centres across the USA, under the sponsorship of the National Institutes of Health (NIH), seeks to collate these findings, find molecular signatures for cancers, and create tumour catalogues.

Reducing all this knowledge of the connections between genes, proteins, and metabolites in a readily assimilable form is a major graphical challenge. The 'metabolic pathways' charts that I was familiar with as a young researcher have become incredibly dense (so dense and complex they are impossible to reproduce here)[16]. Added to them are now gene and protein interaction charts of comparable complexity. So we are already reaping the fruit of the National Science Foundation challenge, even if a full model of the cell is still a long way away.

Biomarkers: tracking disease and response at the molecular level

As we accumulate understanding of the complex system that is human biology, we encounter an ever-increasing number of genetic, proteomic, and metabolomic markers of disease and its response to treatment. Such biomarkers are also the primary tool in personalizing diagnosis and treatment, using them as biological metrics that distinguish one population of patients apparently suffering from the same disease from another.

The literature on this subject has also grown dramatically, and now contains more than 20 000 potential biomarkers.

However, the number of biomarkers recognized by the FDA is much lower, around 170, many of which are markers of potential *toxicity* of drugs. And among those approved are multiple markers associated with individual drugs: warfarin, for example, is contra-indicated for patients with cytochrome 2C9, Protein S and Protein C deficiency, and carriers of the *VKORC1A* allele.

The number of *efficacy* markers associated with new drugs—the basis of stratification of patients—is growing, but from a low base, with cancer being the most important disease area. As we saw earlier, identification of specific gene mutations and/or proteins expressed on the surface of the tumour cell are increasingly important markers for treatment choice. The FDA has encouraged the trend to personalized or precision therapy using such markers. Of the 41 medicines approved in 2014, eight had such markers (half of which were for cancer)[17]. These markers in practical use are, however, a small sub-set of the total potential cancer-related markers in the literature, which include oncogenes, tumour suppressor genes, and non-coding RNAs. We can see that—while there may be challenges moving from biomarker discovery to utilization—we have plenty of raw material from which to craft markers to define diseases and patients at a molecular level.

Cell biology

A number of major developments have revolutionized cell biology. The first, of course, is the discovery and manipulation of stem cells. Whether derived from embryonic cell lines or pluripotent stem cells created from adult cells, they have generated great excitement (an excitement that has often bordered on hype).

The stem cell literature has exploded (see Figure 2.7, a. Stem cell papers). We have been able to differentiate stem cells into a whole variety of cell types. We can now study in vitro the behaviour of different human cells and test drugs on them, for both toxicology and potential efficacy. Clumps of cardiomyocytes (heart muscle cells) beat in a petri dish. Nerve cells conduct impulses. Although these are imperfect models of how the corresponding cells in the body will behave, their human origin make them superior for many purposes to the animal models on which we have depended for so long, and which, of course, carry ethical issues for many.

In the short term, some of the greatest potential lies in screening chemical and biological drugs, but work on the therapeutic applications of the cells themselves is quickening. We will return to this potential later in the book.

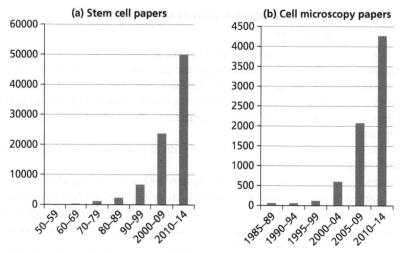

Fig. 2.7 Growth in cell science (up to 2014).
Source: Data from Google Scholar, [search term] 'stem cells'; 'cell microscopy', https://scholar.google.co.uk/, search date 31 Aug. 2015.

We can also study the behaviour of natural human cells with ever greater differentiation and at smaller scale. Cell biology has progressed from study of populations of different cells, to an ability to separate different classes of cells, in (say) the immune system, to focusing on the behaviour and interactions of a single cell. With the benefit of stimulated emission depletion (STED) microscopy[18]—a technique using fluorescence to bypass the normal limit of light microscopy—we can track single cell movements and interactions. This is yet another window on the phenomenon of cancer, giving us clues to how cancer cells invade tissues or move to remote locations to initiate metastases, and how T-cells can locate and destroy them. Cell microscopy has become a major new tool in the biological armoury (see Figure 2.7, b. Cell microscopy papers).

We can also identify and study circulating cancer cells, giving us new tools for assessing the regression of the cancer or its treatment response, and even carry out single cell DNA sequencing

At the sub-cellular level, microscopy can measure the movement of individual labelled proteins within the cell[19]. We can therefore build up from gene to protein to intra-cellular behaviour to cell–cell interaction: a biological continuum.

Synthetic biology

Perhaps most futuristic of all the fields that have emerged in the last 20 years is synthetic biology. In one sense, recombinant DNA—the source of most therapeutic proteins developed in the 1970s and 1980s—is a form of synthetic biology. However, we are now seeing the fusion of molecular biology and engineering in modern synthetic biology in some fascinating new forms. Knowing how DNA codes for proteins and how these proteins interact in pathways to drive metabolism, can we create living cells, or even whole organisms, from basic chemicals?

One of the original decoders of the human genome, Craig Venter, set himself to create a functional, replicating microorganism from a purely synthetic DNA sequence. It required 15 years and a genome of a million base pairs to achieve. Venter's team multiplied the DNA sequence in yeast and then transferred it into a bacterium, which was then able to multiply. The historic announcement of the achievement, in 2010, was made with typical flair. The new microorganisms's genetic code also contained the names of all the authors of the work, plus three quotations, one of them particularly appropriate, from Richard Feynman: 'What I cannot create, I do not understand'.

A little more practically, a team led by Jay Keisling at UC Berkeley has assembled ten genes from three organisms, again in yeast, in an attempt to synthesize the anti-malarial drug Artemensin (currently extracted expensively from the plant *Artemisa annua*).

Synthetic biology will develop organisms to both replicate the production of current biomolecules, and to produce novel proteins 'made to order', which themselves could be therapeutics. However, the majority of drugs to date have been chemical molecules, so let us turn to the disciplines needed to turn our rapidly developing biological insights into practical therapeutics.

Tools for drug discovery

Biology's key partner in the discovery of new pharmaceuticals is medicinal chemistry. Chemical drug discovery has always been the fusion of chemical creativity and biological screening. Advances in both now make the generation of new potential drugs an industrializable process.

Over the last 30 years we have developed automated means of synthesizing potential drugs, using 'combinatorial' approaches[20]. This has dramatically increased the productivity of medicinal chemistry, at least as measured by the number of new chemical molecules synthesized. We have increased the number of drug-like molecules per chemist per year by perhaps 800 times through the 1980s and 1990s[21]. This has led to a massive increase in the sizes of the 'compound libraries' in industry and academia[22].

Keeping pace with the dramatically enhanced ability to generate new molecules as been advances in laboratory technology to enable their high-speed automated testing against assays of activity—so-called 'high throughput screening' (HTS). As a result, this screening process has become 10 times less expensive since the mid 1990s[23]. Therefore we can create and then screen potential drugs at a vastly increased rate and at much lower cost—generating literally millions of options among the 10^{62} drug-like chemical permutations.

In fact, much of the drug design work is now done computationally—'in silico'. Medicinal chemists are able to generate a wide variety of designs and test them for fit in computer models of target proteins before any molecule is ever synthesized. The computational advances that have helped us design drugs are also in use to create in silico models of organs, so that scientists can 'test' the effects of drugs on these models before they ever enter animals or humans. This has proved useful for anticipating cardiac side-effects, one of the most common reasons for dropping promising drug candidates.

So the ability to both produce and test potential new drugs has become dramatically more productive.

New insights from imaging

Imaging technologies have also advanced tremendously. As well as the conventional CT scans to identify a diseased organ or tumour, MRI and PET images can determine the blood flow and level of metabolism in diseased tissues and therefore track the impact of a therapy rapidly and quantitatively (see Box 2.2).

DC-MRI studies, in particular, have seen explosive growth, in investigating an increasing variety of disease states (Figure 2.8).

While imaging equipment is regarded by regulators as a 'medical device' and care needs to be taken in using it, there is obviously greater scrutiny when it comes to implantable devices, to which we now turn.

Implantable devices: biomaterials, mimicry, and miniaturization

We have developed a wide range of materials compatible with human biology. Silicone is probably the best known. Its density is almost identical with that of the human body and it has applicability for much more than breast implants. Titanium bonds well to bone, as we see in dental implants and artificial limb fixation. Spray-on

---

Box 2.2 The example of Dynamic Contrast-enhanced MRI (DC-MRI)

Dynamic MRI (DC-MRI) can distinguish tumours from benign growths by the extent of vascularization around them. It can also monitor the impact of therapy on this blood flow, which is critical for the growth of the tumour. In inflammatory disease, the technology can highlight and measure areas of inflammation, e.g. in arthritic joints, and also shows how these areas shrink and disappear as effective treatment is applied.

The challenge, particularly for hard-pressed radiologists, is to quickly interpret the images, especially the time series produced by DC-MRI. So quantitative algorithms have emerged, such as the DEMRIQ score, that enable them to see a colour-coded, time-dependent series of images that depict blood flow[a]. This technology leads to faster interpretation, greater quantification, and greater reproducibility than simple radiologist readings, and so it is being used increasingly in clinical trials to distinguish the effects of new drugs.

[a] Abstract Archives of the RSNA, 2014 DEMRIQ: A Dynamic Contrast Enhanced MRI Quantification Method for Objective Assessment of Treatment Response in Patients with Inflammatory Arthritis.

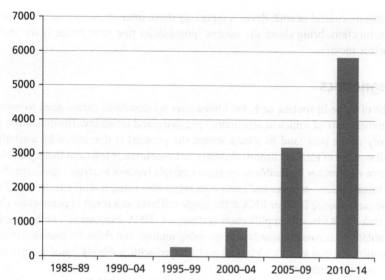

Fig. 2.8 Growth in the application of DC-MRI (up to 2014).
Source: Data from Google Scholar, [search term] 'DC-MRI', https://scholar.google.co.uk/, search date 31 Aug. 2015.

medical adhesives, based on those developed for industrial purposes, can be used to temporarily seal major injuries or replace sutures in small surgical wounds. The new technology of 3D printing has enabled the manufacture of very precise and customized implants, and shape memory has been incorporated in implants using magnetic materials.

In addition, biodegradable materials have been developed in large libraries for drug, cell, and gene delivery applications[24]. In many cases, scaffolds made of biomaterials are combined with cells to replace or enhance joints. Accordingly, the literature dealing with biomaterials has grown dramatically.

The last 20 years have seen some major possibilities opened up for the design of medical devices, employing these materials. We have also developed a more detailed understanding of electrophysiology, especially of the heart, opening up the possibility of a range of devices for modifying and restarting heart rhythms.

Understanding of how pain and sensation is conducted has made possible electro-stimulatory devices in the brain or peripheral nervous system, with potential for how artificial limbs could function.

Technology for minimally invasive surgery has developed via a host of miniature devices, from cameras to liquidizers. Surgeons can now operate through a 'keyhole' or through arterial entry routes. And, of course, telemedicine has made it possible for surgeons remote from the patient to take control of the operation.

Linking devices to diagnostics opens up the possibility of therapy dispensers (like the insulin pump) that can continuously and automatically administer a drug in response to biological signals.

Software control of such devices, especially those intervening in cardiac control or brain function, bring closer the 'bioinic' possibilities that were formerly the stuff of futuristic movies.

Diagnostics

To be of value in routine care, the biomarkers we described earlier need to become diagnostic tests of sufficient sensitivity, specificity, and reliability. The diagnostic laboratory of the past (and to a large extent the present) is dominated by a relatively small number of routine clinical chemistry or immunoassay tests and tissue pathology. Science now makes it possible to measure multiple types of analytes—genes, proteins, and metabolities—at lower and lower concentrations, using multiple assay modalities.

We can measure DNA or RNA at the single cell level, as a result of polymerase chain reaction (PCR)-based amplification techniques. DNA expression can be assessed quantitatively across thousands of genes using multiplexed chips. Proteins at nanomolar concentrations can be identified and assayed with antibody-based arrays, or in mass spectrometers.

Most of these techniques start as labour-intensive one-off laboratory experiments. However, we are now adept at placing them on a wide variety of instrument platforms for high volume application. These instrument systems automatically handle samples, programme the specific tests needed on each specific sample ('random access'), and route the samples to the appropriate instruments. The results can then be routed through a laboratory information system automatically to the treating physician.

So the technology is there for most of the biomarker types we may want to turn into routine tests.

The data revolution

Technology has advanced not only in the laboratory and the clinic but also in real-time tracking of treatment effectiveness in patients, whether in clinical trials or routine care. Wireless technology, and mobile devices have the potential to speed the collection of results in clinical trials.

Now that many patients have smart phones, apps loaded on them can record patient-reported outcomes (PROMs). Social networks also (inadvertently) collect information on the effectiveness or side-effects of treatments, as sufferers exchange experiences, for example via their Facebook networks. Membership of virtual organizations, such as PatientsLikeMe, enables patients to input and share their own views of efficacy, side-effects, or convenience. Artificial intelligence systems can analyse such (anonymized) flows of experience. We can therefore identify where and for whom the treatment is working—and when, for whom, and to what extent unexpected side-effects emerge as more and more patients have access to it.

The other record of patient experience, through the eyes of the clinician, is the electronic health record, becoming standard in most parts of the developed world. While there are still problems with individual systems communicating with each other, particularly between those in primary and secondary care, the data is a potential treasure trove. Patients can be screened for suitability for clinical trials. Once a product is approved and is widely available, correlations can be established between patients' responses to treatment and other factors in their 'health lives', and signals on rare side-effects can be separated out from the background noise.

So the same information technology that we discussed in Chapter 1 and which has led to many revolutions in our lives stands ready to do the same in healthcare.

Our scan of the science and technology developed over the last 20 years should give us heart that the potential is there for multiple new starting points for innovation. Many of these innovations also have the potential to disrupt and transform the innovation process itself.

Another powerful consequence of many of these tools and technologies, one we will explore in more detail later, is their ability to 'see' much finer molecular and structural detail. This creates a greater ability to personalize diagnosis and treatment, dissecting disease on a patient-by-patient basis as never before. So 'personalized' or 'precision medicine' should be one of the most potent themes of the new biology. Disease states that were formerly bundled together because of common symptoms, or a common organ at which they appear, break down into different diseases when studied at the molecular level. Alternatively, commonalities emerge between diseases with different patterns of symptoms.

Cancer, which has in many respects led the personalization revolution, is now being reclassified in molecular rather than organ-of-origin terms. We are also realizing that tumours are made up of many different colonies of cells and that individual cells have different mutation patterns. While this makes developing therapeutic strategies more complex, it begins to explain the wide range of responses that patients have to the same treatment regimen. Their tumours are very different, even if they look the same to the radiologist or pathologist.

A widening global research effort

For most of the last 20 years, bioscience research has been dominated by the USA, Europe, and, to a lesser extent, Japan. However, there are now new national forces emerging, investing heavily in medical innovation. China is a rapidly growing presence in bioscience. Annual publications in the field from Chinese researchers have grown from 115 000 in 2009 to 155 000 in 2013. Of the approximately 1.7 million papers published in science, technology, and bioscience in 2013, a third came from the EU, 20% from the USA, and nearly 40% from Asia, with China the leading contributor. China has already surpassed the UK, and is forecast to overtake the USA as a producer of academic papers by 2020![25]

So medical innovation is now a global pursuit, with the prizes—from Nobel awards to new science-based businesses—likely to be distributed more widely in the future than in the past. But there will only be prizes—for the patients and for innovative businesses—if we confront the downstream difficulties that defeat innovators, globally.

In this chapter, I have done scant justice to the quality and content of the research in each of these explosively growing individual fields. However, I have hopefully done enough to convince the reader of the rapidly growing *quantity* of activity and publications, and therefore possibilities in the field over the last 20 years. The challenge that has faced life science translation has certainly not been the result of lack of raw material. In fact, the scientific efforts and literature are now at a scale at which it is impossible for researchers to keep current in even quite narrow fields. The total body of academic output in life sciences has grown steadily and provides a backdrop for the much more explosive growth of the specific fields highlighted in this chapter (Figure 2.9). It also serves to illustrate how impossible the task has become to stay current.

In Chapter 3, we analyse how effectively the current innovation process has translated this research output into practical products and patient benefit. However, in *very* broad terms, Figure 2.10 gives a sense of the current level of activity in life sciences, on a logarithmic scale, at the various stages of development, with pharmaceuticals as the example. The fall-off, from research input to usable output, is truly dramatic.

The precise numbers are not important: they are broad estimates, with no attempt to adjust for the typical time lag between stages. However, we see how modest is the ultimate output for the scale of activity at earlier stages. Box 2.3 outlines how these estimates were made.

In Chapter 3, we will go on to deconstruct the gaps that exist between these various stages in the innovation process.

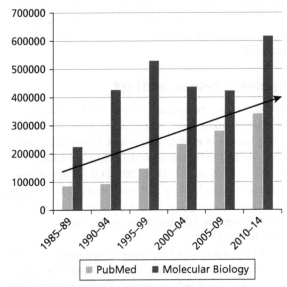

Fig. 2.9 Growth in biology-related scientific literature (up to 2014).
Source: Data from Google Scholar, [search term] 'PubMed'; 'molecular biology', https://scholar.google.co.uk/, search date 31 Aug. 2015.

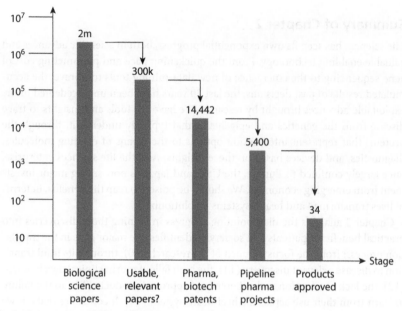

Fig. 2.10 Measures of life science output.

Box 2.3 Life science metrics

Around 1.9 million papers in life sciences were published in 2014 (2010 estimate by Stanford University research, quoted in www.quora.com, scaled by 5% growth per year to 2014), of which only around 15% (300 000) might be estimated to be reliable, replicable, and relevant, consistent with academic analysis on 'waste' in research[a]. In 2014, 14 442 Patent Cooperation Treaty (PCT) patents in the field were filed, according to the World Intellectual Property Organization[b]. The estimated 5400 products in the pharmaceutical pipeline in 2012, sourced from PhRMA (the US pharmaceutical association), is a reasonable measure of development activity (although it may contain some duplications). The 34 products that emerge, an average of the last 4 years of FDA approvals of new molecular entities, is the most precise and reliable number of all[c], and also, by two orders of magnitude, the smallest, and perhaps only one-third of these go on to widespread use for patient benefit.

[a] Glasziou P, Altman DG, Bossuyt P, et al. 2014. Reducing waste from incomplete or unusable reports of biomedical research. *The Lancet*, 383: 267–76.
[b] WIPO, quoted in www.patentdocs.org/wipo-releases-data-on-2014-international-patent-filings
[c] FDA, 'Development & Approval Process (Drugs)', http://www.fda.gov/Drugs/DevelopmentApprovalProcess/, accessed 15 Jul. 2016.

Summary of Chapter 2

Life sciences has seen its own exponential progress, both in scientific advances and valuable enabling technology. From the quickening pace and plummeting cost of gene sequencing to the emergence of new data-mining tools to unravel the accumulated results of past decisions, the last 20 years have been unprecedented in the exploitable advances brought by research. We have the tools and insights to trace disease from the genetics and epigenetics that typically underlie it, through the proteins that represent intervention options to the means of creating molecules, diagnostics, and devices based on those insights. And the life sciences enterprise, once largely confined to Europe, the USA, and Japan is now seeing major investment from emerging economies. We should be poised to reap the benefits, in terms of lives transformed and health systems revolutionized.

Chapter 3 analyses the disappointing progress in turning these discoveries into practical benefit for patients and society. It identifies five major gaps in the translation process: from the focus of much of the research (T0), through its fitful transition to disease-relevant discovery (T1), the high fallout rate in clinical development (T2), the lack of adoption and adherence of approved products (T3), to the failure to learn from their use across the healthcare system (T4). It concludes that—if we are to reap the harvest of new products and medical advance from the scientific investments just described—a major overhaul is needed of the translation process itself.

References

1. **Sanger F.** 1988. Sequences, sequences, and sequences. *Annu Rev Biochem*, 57:1–28

2. **Powledge TM.** 2003. How many genomes are enough? *The Scientist*, (Nov 17, 2003)

3. **Blackman G.** 2010. Gene genie. *Sci Comput World*, 113:12–14.

4. **Du Z, Fei T, Verhaak RG,** et al. 2013. Integrative genomic analyses reveal clinically relevant long noncoding RNAs in human cancer. *Nat Struct Mol Biol*, 20:908–13

5. **Kellis M, Wold B, Snyder MP,** et al. 2014. Defining functional DNA elements in the human genome. *Proc Natl Acad Sci U S A*, 111: 6131–8.

6. **Khoury MJ, Valdez R, Albright A.** 2008. Public health genomics approach to type 2 diabetes. *Diabetes*, 57: 2911–14.

7. **Herman JG, Latif, F, Weng Y,** et al. 1994. Silencing of the VHL tumor-suppressor gene by DNA methylation in renal carcinoma. *Proc Natl Acad Sci U S A*, 91:9700–4.

8. **Mendizabal I, Keller, TE, Zeng J, Yi SV.** Epigenetics and evolution. *Int Comp Biol*, 54:31–42

9. **Brand D.** 2004. From Years to Hours: Calculating Molecular Structure Speeds Up, Fueled by New X-ray Sources. Ithaca, New York: Cornell University. Available from: http://www.news.cornell.edu/stories/2004/02/new-x-ray-sources-speed-protein-crystallography, accessed 01 Sep. 2015

10. **Van Brunt J.** 1986. Protein architecture: designing from the ground up. *Nat Biotechnol*, 4:277–83.

11. Membrane Proteins of Known 3D Structures, Stephen White laboratory at UC Irvine, http://blanco.biomol.uci.edu/mpstruc/ accessed 01 Sep. 2015

12. National Center for Biotechnology Information, US National Library of Medicine, http://www.ncbi.nlm.nih.gov/structure/ accessed 01 Sep. 2015

13. The Structural Genomics Consortium, Oxford, Toronto, www.thesgc.org/structures, accessed 01 Sep. 2015

14. The Structural Genomics Consortium, Oxford, Toronto, http://www.thesgc.org/scientists/resources/kinases, accessed 01 Sep. 2015

15. **Dhillon AS, Hagan S, Rath O, Kolch W.** MAP kinase signalling pathways in cancer. *Oncogene* 26:3279–90.

16. Hoffmann-la Roche Ltd, Biochemical Pathways, http://www.roche.com/sustainability/for_communities_and_environment/philanthropy/science_education/pathways.htm, accessed 01 Sep. 2015

17. US Food & Drug Administration, http://blogs.fda.gov/fdavoice/index.php/tag/biomarkers/, accessed 01 Sep. 2015

18. **Eggeling C, Willig KI, Barrantes FJ.** 2013. STED microscopy of living cells—New frontiers in membrane and neurobiology. *J Neurochem*, 126:203–12

19. **Muzzey D, van Oudenaarden A.** 2009. Quantitative time-lapse fluorescence microscopy in single cells. *Annu Rev Cell Dev Biol*; 25:301–27.

20. **Geysen HM, Schoenen F, Wagner D, Wagner R.** 2003. Combinatorial compound libraries for drug discovery: an ongoing challenge. *Nat Rev Drug Discov*, 2:222–30.

21. 2000. Combinatorial chemistry. *Nat Biotechnol*, 18:Suppl:IT50–52.

22. **Dolle RE.** 2011. Historical overview of chemical library design. In: Zhou JZ, ed. *Chemical Library Design (Methods in Molecular Biology 685)*. New York: Springer Science, pp. 3–25.

23. **Mayr LM, Fuerst P.** 2008. The future of high-throughput screening. *J Biomol Screen*, 13:443–8.

24. **Anderson DG, Levenberg S, Langer R.** 2004. Nanoliter-scale synthesis of arrayed biomaterials and application to human embryonic stem cells. *Nat Biotechnol*, 22:863–6.

25. **Jang H, Kim H.** 2014. Research output of science, technology and bioscience publications in Asia. *Science Editing*, 1: 62–70. Available from:http://escienceediting.org/journal/view.php?number=16. Figure: http://escienceediting.org/upload/thumbnails//se-1-2-62f1.gif

Chapter 3

The gaps in translating biomedical advance into patient benefit

While scientific advance has been unprecedented over the last two decades, the output of useful products and patient impact has not kept pace. Indeed in some areas, like pharmaceuticals, we have seen continuation of a long-established downward trend in productivity, or output per unit of input. Not only that, despite technologies that have the potential to speed the translation process, the time lag between research discovery and patient benefit has remained around 10–15 years in many areas, and in some cases has risen.

Although we know much more about basic biology, we are little further forward in understanding the mechanism of many diseases. We have industrialized many of the steps in generating and screening potential products, but somehow the actual numbers of clinical candidates that emerge has increased only slowly, and this increase has been focused in a relatively few areas, such as cancer and immunology. Most new products still alleviate symptoms or slow the course of disease rather than reverse it.

Clinical trials have seen increased levels of attrition, especially at the most costly, later stages. Regulatory requirements have mounted, and the pressure on health systems has generated a new barrier that products must surmount—HTA. At the prices that companies feel they need to charge to sustain returns as R&D productivity declines, this hurdle gets higher each year.

Once products reach the market, innovators still rely on conferences, journals, and medical representatives to spread the word, and this still takes up to a decade. Even once they are prescribed, most drugs face one final barrier to delivering health improvement—patients that are unclear or unpersuaded about their benefit and therefore prone to give up using them.

So almost all the stages in the process are 'broken' to one degree or another. We need to understand how and why, at each stage, in order to design reform. Some of these factors will lead us to mechanistic changes in the process. Others point to cultural gaps between academia, small start-up companies, large international majors, government agencies, patients, and the health systems that ultimately must pay the bills. These are tougher to change, as they are based on some stark differences in perspective on what represents value, and what level of risk is acceptable.

If we are to overcome the challenges we need radical change, in which these groups must all engage, individually and in collaboration. This will be the subject of Chapter 5. At some points, however, we already see the beginnings of this change process, but it is often at a painfully slow pace, somehow out of sync with the speed at which most other facets of 21st century society are moving. If we do not quicken the pace, we risk frustration from politicians, investors, and most of all patients.

Gaps in translation

There are three well-recognized 'gaps in translation' in medical innovation science, termed **T1**, **T2**, and **T3** by experts such as Tufts Clinical and Translational Science Institute[1]: **T1** is the gap between a research discovery that promises practical medical advance and a potential therapy that is ready for clinical trial. **T2** is the gap between candidate products undergoing trial and 'approved' products with regulatory clearance for, and reimbursement by the health system. **T3** refers to the gap between the products that are available and those that are widely adopted and properly used.

This is a good framework for much of our analysis, but I would add two new 'gaps' at either end of this process:

> **T0**—the knowledge gap that exists between our understanding of fundamental biology and our knowledge of the specific changes that drive disease. We see a good example of a T0 gap in motor neuron disease or ALS: while we know that the patient's motor neurons die step-by-step over the period of a few years, we don't know why. And without a 'why', it's hard for drug discovery to be more than random shots in the dark.

> **T4**—the gap between the millions of treatments delivered daily and the knowledge of outcomes—what really works and how it could work better. Until very recently, we had little systematically collected information about the link—or lack of it—between treatment choice and health impact. With the advent of electronic health records and databases, we now know that some treatments we thought were sound (like beta-blockers for congestive heart disease) do more harm than good, while other treatments (like metformin for diabetes) in fact deliver benefit elsewhere, in this case in protection from ovarian cancer.

So my expanded model of medical innovation science, and the gaps in our current capabilities, looks like the scheme shown in Figure 3.1.

Let us now look at each of these five gaps in greater detail, to discover their basic causes. Without this understanding, actions and policies to bridge them will themselves be shots in the dark.

T0: do we know what underlying biological problem we need to solve, when we address a disease?

Disease pathology is much less straightforward than it sounds to the layman. Bound up with this issue is that of the diversity of genetic causes, behavioural traits, and metabolic changes that can result in similar symptoms, and the heterogeneity of many patient populations.

The search for a single biological pathway and specific intervention points has sometimes been highly successful: we will see some examples in Chapter 4. However, many of the disease targets we now address are much more complex than the blood pressure and lipid control, or stomach ulcer intervention that have been triumphs of the last 20 years. They are degenerative disorders that involve multiple genes, proteins, and pathways.

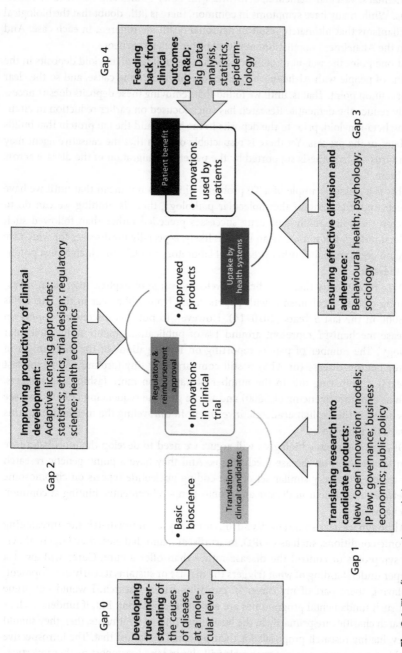

Fig. 3.1 The five gaps in translation.

Dementia is a case in point. Firstly, we need to sub-segment it into 'classic' Alzheimer's, vascular dementia, dementia with Lewy bodies and other, rarer conditions. While many have symptoms in common, there is little doubt that the biological mechanisms that ultimately result in neuronal death are different in each case. And even the Alzheimer's population may itself be quite heterogeneous.

At one point the weight of opinion was that the buildup of amyloid deposits in the brains of people with Alzheimer's disease was the underlying cause, and so the clear intervention point. That is, until we found that removing these deposits doesn't necessarily reduce the dementia. Research has since focused on earlier reduction in circulating beta-amyloid, prior to the deposits in the brain, and the tau protein that builds up later in the process. Yet there is respectable opinion that the causative agent may be a virus—a hypothesis supported by the pattern of migration of the disease across the brain.

This is a classic example of a T0 problem. This does not mean that until we have a deep understanding of the molecular pathology, there is nothing we can do to develop therapies: many past drug successes preceded rather than followed such understanding. But it does mean it is less likely, especially for diseases for which we have no good animal or other preclinical laboratory models on which to test potential therapies.

Over recent years there has been a welcome surge of explicit interest in investigating disease mechanisms, with papers rising from 200 a year in the late 1990s to 8000 in the last 5 years (2010–14). However, to put this in context, papers on 'disease mechanism' represent around 1% of publications mentioning 'molecular biology'. The number of papers reporting on investigations of the mechanism of motor neuron disease (or ALS) is still comparatively tiny (around 650 in the last 5 years). Comparing this to the number of papers on more fashionable subjects, such as exome sequencing (17 800) and whole genome sequencing (19 100) we see how small is the explicit academic investment in unraveling the mechanism of this devastating disease.

Of course, the basic biologists will argue: we need to develop the tools before we turn to use them on disease mechanisms. And they have a point: genetic research has uncovered a very similar set of non-coding nucleotide repeats on chromosome 9 between motor neuron disease and fronto-temporal dementia, hinting at common mechanisms[2].

Although we have some effective drugs for many less immediately life-threatening chronic conditions, such as COPD, heart disease, and diabetes, they largely alleviate symptoms or control the disease rather than offer a cure. Cures will await a deeper understanding of what triggers the disease or sustains its early development.

Having spent part of my career in curiosity-driven research, I would not argue that such fundamental programmes are not vitally important. But, if funders such as research charities are primarily in the business of finding treatments, then they should be evaluating research proposals for their potential to do just that. The introspective world of peer review of grant proposals will always tend to support further investigation of fundamentals.

T1: why does so little promising innovation emerge from the academic lab?

It is said that 'what happens in Las Vegas stays in Las Vegas'. Or at least it did before Facebook came along. But the same can sadly be said of much that happens in even the most innovative academic labs. The ideas and experimental results escape in the form of peer-reviewed papers, of course, or the researchers would soon be looking for other positions. But very few of the findings published in the ever-spiralling biomedical literature we saw in Chapter 2 are translated into products for human testing. Why might that be?

The answer is not simple, and has scientific, technical, organizational, and cultural components. Scientifically, the practical steps towards a real product are not attractive to most academics, nor professionally rewarding. Technically, few academic labs have the prototyping or early development skills or facilities needed. Organizationally, many universities are not set up to collaborate intimately with the companies that do have those skills. Culturally, the whole 'commercial' enterprise is seen by many academics as slightly suspect and certainly less interesting than their normal pursuits.

Scientifically, the interest of most academic scientists runs out long before a commercializable product is reached. Their career-making *Nature* or *New England Journal of Medicine* papers, reviewed by their academic peers, illuminate the mechanisms of normal biology and—less frequently—of disease, and often pinpoint genes, proteins, pathways, and metabolites of significance found in their experiments. These, as we shall see in the Gleevec story in Chapter 4, simply represent starting points for the development of effective products. Developing these molecules themselves requires close collaboration with medicinal chemists, professional drug designers, of which there is not a ready supply in academia.

In any case, the *technical* task of designing a drug (or a device for that matter) is one of multiplexing and progressive iteration. In the drug sphere, the initial finding that a 'molecular probe' binds to a relevant target site is the beginning not the end of a journey, often one of several years' duration. Once a promising class of molecules is found to target the right receptor site, the binding constant has to be boosted; then the selectivity (the ability to bind where needed but not where it will cause side-effects) needs to be worked on. Once a potent and selective structure is found, a dosage form needs to be developed that can—either orally or by infusion—be used in patients, i.e. has a good set of pharmacokinetic and pharmacodynamic (PK/PD) properties. (PK/PD properties define whether the drug reaches the organ of interest in sufficient concentration and is then broken down by the body to avoid toxic build-up.)

These are all very scientifically challenging problems, but most academic researchers do not have the interest, skills, or resources to engage in them. Those with that interest and those skills and resources tend to be in commercial companies. Some academics do value access to products or potential products, in order to probe their hypotheses further: that is one of the strengths of public–private collaborations like the SCG mentioned in Chapter 2 (see 'Proteins—mapping their structure and function').

This gap between initial findings and a useful product has not stopped researchers and universities from patenting their discoveries: universities were the holders of a quarter of the biomedical utility patents by the mid-1990s[3].

This patenting, though understandable, also contributes to *organizational* barriers. Technology transfer offices of cash-hungry universities have been very active in bioscience, although sometimes misguidedly. They frequently seek to maximize royalties from licensing rather than realizing that close academic/industry collaboration is what is needed to take most life science discoveries forward. When there does appear to be commercial potential, spin-out companies are often created, but most lack the critical mass to be sustainable. Large commercial companies, of which there are perhaps 30 of global scale, simply cannot engage with all the potential products that could be developed from academic research. Spin-out companies, with very limited financial resources, attempt to take these forward, but too many of these fall at an early hurdle, due to inexperience or lack of cash.

For all these reasons, only a small proportion of academic discoveries spawn commercial programmes. But could or should they?

One of the most concerning discoveries of recent years is the sheer irreproducibility of the results of many academic researchers. Depending on the field, around half of the published results cannot be reproduced by other laboratories. Very rarely is this a result of deliberate fraud, or of such negligence as was revealed in the recent claim that induced pluripotent stem cells could be tricked into differentiating by simply adding citric acid—a result that swept the cell therapy world before being found unreliable.

More often the publication of results before they are conclusively confirmed result from the rush to 'publish or perish'. With the ability now to claim precedence by means of online journals that publish within days, the incentive to conduct confirmatory experiments reduces still further.

A further factor is the inherently sensitive nature of some biological experiments, especially those in cell systems. I have personally observed this in a programme seeking to create hepatocyte-like cells from EPS cell lines for toxicity testing. Even the same university laboratory produced cells with different characteristics at different times, with different researchers involved.

As a result, commercial companies now approach academic findings with great skepticism. Most large companies routinely seek to reproduce the experimental result in their own laboratories before investing any time in seriously pursuing it. Some estimates of the proportion of academic results that are irreproducible are as high as 70%[4,5].

A further factor in non-translation is the focus of much academic research. For some scientists, the ability to do an experiment in a simple model system, and then be able to present results that interest their peers, outweighs consideration of whether the experiments really have any relevance to human health. Studies of mice with 'dementia' come to mind. Such work may not be translated from bench to bedside because it is fundamentally untranslatable.

There are also *cultural* gaps between the two worlds of academic science and commercial product development. Too few individuals have spent time in both worlds, and stereotypes of each other tend to dominate thinking on both sides.

T1 yield

Can we analyse the trend in outputs over inputs across this second gap—what we might call 'T1 yield'? Steady increases in crude input measures, such as NIH funding or papers in PubMed, have not been reflected in numbers of candidate products. One measure of the 'innovation yield' across T1 is based on citation-weighted patent output per unit of R&D investment, which fell by two-thirds over the 2000s[6]. There is a major and growing gap in translation at the bench-to-candidate-medicine level.

Recalling the enormous growth in research on new tools and technologies that we reviewed in Chapter 2, the decline in T1 yield is clearly a paradox. With DNA sequencing efficiency increasing by around 2 billion times, 10 000 times fewer man-hours required to compute protein 3D structures, and roughly 800 times more drug molecules synthesized per chemist, our ability to translate research into products should have advanced, not retreated.

The problem does not seem quite as serious in *medical devices*, or at least it is harder to measure. The T1 process in devices is typically quite different: the majority of prototype devices come not from a bench scientist but from a surgeon or interventional cardiologist improvising to solve a problem in the operating theatre. I am privileged to chair a Bright Ideas fund committee at Guys & St Thomas's Hospital in London and see a range of medical device proposals to, for example, maintain patients' mobility when dependent on direct heart support, or circumvent catheter attachment problems. There seems no shortage of ideas or enthusiasm to translate them; it is usually early stage funding that is the main barrier.

So how many possible devices aren't invented that might be?—much harder to say. However, we will see that those that are invented are not guaranteed a smooth passage to commercialization.

Diagnostics, however, clearly face a problem at the discovery-into-development transition. Many of the large number (perhaps 20 000) of biomarkers in the biomedical literature could in principle help in the targeting of medicines. Of these, only 170 have been incorporated into FDA approvals or guidance to date. Most diagnostic companies do not have the resources to mount collaborative programmes with academic researchers.

T2: why does it take many years and many billions to produce every new product?

The spiralling cost and lengthy timescale of biomedical development, especially of drugs, is now well-documented. In the pharmaceutical industry, it was sometimes worn as a badge of honour: 'Look how hard this is, how much we have to spend, how few products make it through to market.' A wiser approach would have been to ask: Why? And how can we radically reduce time, cost and attrition rate? Everyone now involved realizes the overall pattern is unsustainable. Every individual investment made in the pharmaceutical industry is, in a sense, a bet that the odds can be beaten.

Output measures are easier to come by at the T2 stage. We know how many drugs are approved versus, for example, patents filed, the number of drugs in development at various stages, or clinical trial expenditure.

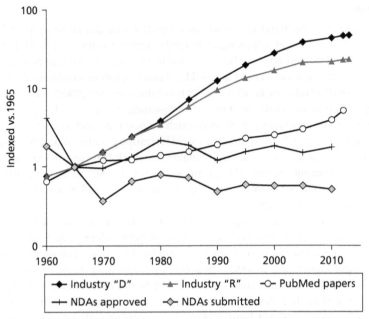

Fig. 3.2 R&D input and output metrics.
Reproduced with permission from Barker RW and Scannell JW. The Life Sciences Translational Challenge: The European Perspective. *Therapeutic Innovation and Regulatory Science*, Volume 49, Issue 3, pp. 415–424. Copyright © 2015 Develop Innovate Advance (DIA).

Until very recently—and over a remarkably long period—growth in inputs in terms of patents and drug development expenditure substantially outstripped growth in the output in terms of drugs approved[6].

Figure 3.2, with its logarithmic scale, shows how much faster development expenditure has climbed versus new drug approvals (NDAs) over a 50-year period.

This input/output imbalance is clearly not sustainable. In the past, companies have relied on their periodic ability to launch successful 'blockbusters' (drugs selling more than $1bn per year) at high prices. This has enabled them to recoup their R&D expenditure for both successful and unsuccessful programmes. This has worked for some companies: others, less successful in converting their pipelines into products, have been acquired—for the products from past research—with the combinations justified financially by infrastructure rationalization across the merging organizations.

So the most important factor contributing to the overall R&D cost spiral is attrition—the steadily increasing proportion of products that fail in clinical trials to prove that they are sufficiently effective, or adequately safe, or both. There can be no other industry on the planet in which as few as 5% of the products that are seriously trialled make it through to routine use.

The traditional process of clinical development is depicted in Figure 3.3. Phase 1 studies, typically in human volunteers, test safety. Phase 2 studies find optimal dosages

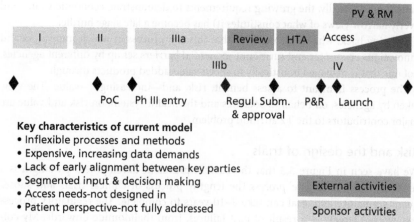

Key characteristics of current model
- Inflexible processes and methods
- Expensive, increasing data demands
- Lack of early alignment between key parties
- Segmented input & decision making
- Access needs-not designed in
- Patient perspective-not fully addressed

Fig. 3.3 Current development path.

and gain the first evidence of efficacy, the 'proof of concept'. Phase 3, the 'pivotal trial' (or often trials) demonstrate that the overall benefit/risk ratio is positive. The most serious attrition occurs in the Phase 2 and Phase 3 trials, the latter often costing tens or hundreds of millions of dollars to conduct.

The most worrying of the attrition trends is the fall in the success rate in Phase 3, with the trials at this stage frequently costing tens or hundreds of millions of dollars. Even more frustratingly, for innovators and patients, are cases in which drugs make it through the regulatory hurdle, only to be rejected or severely restricted by HTA agencies or payers.

Some point to these levels of attrition and associated huge investment per new product as an entry barrier for small entrepreneurial companies that is tolerated, or even welcomed, by large companies. But it seems unlikely that any company, large or small, welcomes a situation in which 'T2 yield' has declined to the point that such a tiny percentage of candidate drugs entering Phase 1 ultimately gain approval.

Breaking this down, two-thirds of drugs fail on safety in Phase 1; of those that continue, 60% fail in Phase 2, typically on lack of efficacy, and two-thirds of those that enter Phase 3 fail to demonstrate adequate benefit/risk. The wastage in investment, in human effort by investigators, and in willing patient participation is staggering.

There are many important factors that contribute to this poor T2 yield. These include overly rigid development models, often followed by companies in a 'safety-first' manner, not wishing to risk cutting corners with a valuable asset. The ever-increasing demands of the regulatory system, leading to the increasing burden of ever-larger clinical trials, is a second factor. The traditional approaches to clinical trial design and evaluation, and the way in which knowledge and responsibility is siloed along the development process compounds the problem. There are cultural and organizational barriers to collaboration between companies researching similar products. Difficult to quantify but I believe also important is the widespread exclusion from the R&D process of the patient's perspective and the consequent 'one size fits all' approach to patient need. Finally, the growing requirements to demonstrate economic value (and overly narrow views of what constitutes it) has become a late stage hurdle.

Stepping back, what is also clear is that this is a process in which innovators and innovations encounter and surmount sequential barriers set up by different agencies, not one that is managed holistically to speed value-added products through.

The process is meant to assess benefit, risk, and—increasingly—value. The view taken by agencies, on behalf of societies and the health system, on risk and value are major contributors to the T2 attrition problem.

Risk and the design of trials

We have seen in Figure 3.3 that the current prevalent development paradigm is a relatively *rigid 'stage-gate' process*.The lengthy process of clinical trials and subsequent regulatory appraisal can take 8–10 years to complete. The regulatory process has evolved largely as a result of past failures, from thalidomide onwards. My colleague Peter Lachmann calls the evolution of drug regulation 'The Penumbra of Thalidomide'—a long shadow cast by a single tragic episode in pharmaceutical innovation (one we shall consider in Chapter 4).

Following thalidomide, specific tests were introduced to ensure that no future drug would cause the same kinds of developmental abnormalities. But it also reinforced the philosophy that regulation must steadily develop to ensure any significant problems are never repeated once a product goes into wide use in a general population.

In response to these regulatory demands, the *scale and cost burden of clinical trials* have escalated. We can see the scale of the change that has resulted, if we compare: (a) the first randomized controlled trial (RCT), published in 1948, which randomized just 107 patients; (2) the number of patients per pivotal trial for antihypertensive agents, that rose from around 200 to around 450, between 1987 and 2001; (3) the average number of patients across pivotal trials for oral anti-diabetics, that rose from around 900 to over 4000 between 1993 and 2006[6].

Looking at this phenomenon for just a single type of agent, the first 1994 pivotal trial for Merck's lipid-lowering drug, simvastatin, recruited around 4400 patients, while its trial for anacetrapib, a potential addition to lipid-lowering therapy, is currently recruiting 30 000 patients.

Much of this results from a 'regulatory ratchet' that decrees that, if a class of product demonstrates a specific major safety problem once approved, other products for the same patient population are put through new, or more rigorous tests. Regulatory steps are rarely removed, of course. Some elements of the process—notably some forms of animal testing—are believed by many to have poor predictive value, yet regulators are not enabled to drop them for fear of criticism or some future disaster that they might have prevented.

The avoidance of rare, serious side-effects is a major preoccupation of regulators, for perfectly proper reasons. However, they cannot be eliminated even by substantial increases in Phase 3 clinical trial sizes, both for statistical reasons and because trial populations are unrepresentative of the ultimate recipients, who may have other diseases or propensities to develop them.

The underlying philosophy of the existing model is that innovators should bear all the burden, cost, and risk for the development process, and then receive all the financial benefit. The regulatory bodies act as gatekeepers or watchdogs at the end of the process, ensuring that the companies do not try to launch defective products that will cause damaging side-effects. Funding of the regulators by industry (through, for example, the regular FDA-industry PDUFA [Prescription Drug User Fee Act] nego-tiations, and similar sponsor-funding arrangements in Europe) must been carefully scrutinized, to avoid any implication that industry funding undermines the objectivity of the watchdogs.

This 'confrontational' philosophy has evolved somewhat, with regulators being pre-pared to offer 'scientific advice' to innovators at earlier stages. However, the idea that both innovators *and* regulators should bear some responsibility for the declining pro-ductivity of the innovation process is beginning to take root. There have therefore been a number of recent improvements or promising pilots introduced by regulatory bodies becoming conscious of their responsibilities not only to catch dangerous products but also speed high-value innovations to the patient, and we will return to these later.

The *design of clinical trials* has also ossified. The concept of the clinical RCT has, since its introduction in 1948, become the 'gold standard' of clinical evidence. Patients are randomly assigned to the new treatment or to either placebo or the standard pro-tocol, without either patient or doctor knowing which is which, and the results com-pared. Some regulators and also HTA agencies will accept no other standard. However, senior regulators such as Mike Rawlins have begun to question this publicly, especially where it is either impractical or unethical to run such a trial[7]. There are several alterna-tives we will consider later.

Central to the RCT is the principle of 'equipoise': that there is no a priori reason to believe that the treatment under study is superior to the current standard of care (or placebo) given to the patients in the control arm. Therefore patients in the control arm ought not automatically be disadvantaged by that assignment. This is, however, patently not the case when the standard treatment is known to be ineffective and there is some evidence of efficacy in the new product. (See Box 3.1.)

In such situations, at a minimum the patients in the control arm ought to be offered 'crossover' to the new treatment if the trial begins to prove efficacy. In Chapter 4 we will address other approaches to this problem.

Having reviewed the technical challenges in the historical approach to development and regulation, let us now turn to some of the attitudes that underpin them—attitudes to risk itself, to the role of the patient, and to the role of research in the health system.

We are living in a society in which risks must be highlighted and avoided, some-times, it seems, at all costs. The precautionary principle reigns. (I have just assembled a barbeque grill bearing the warning that it carries a risk of burning!) Unfortunately, all drugs carry risk: references to 'safe' drugs, or devices, are totally out of place. 'Acceptably safe' is all we can and should say. The development process is an exercise in determining benefit/risk (and also increasingly value/cost) in the face of a number of inherent challenges.

Firstly, the limitations of RCTs, in terms of the type and amount of data they yield. Patients are highly selected to be free of any co-morbidities and readily accessible to

Box 3.1 The case of clinical trials in metastatic melanoma

Metastatic melanoma has been an almost invariably fatal disease, in a short space of time. Traditional chemotherapy offers little life extension. We are now seeing a generation of much more promising agents—that seem able to extend life for some patients for several years—and, of course, patients are anxious to join clinical trials. However, many of them discover they have been randomized to the control arm: they point out that this is not really equipoise—it is a potential death sentence. Others, excluded from the trials altogether because of strict inclusions and exclusion criteria, describe applying traditional rules in this situation as cruel 'medieval torture', in the cause of the science of the RCT.

This issue, and the need to reform clinical trial design in such situations, has been highlighted by Bettina Ryll, who lost her husband to the disease and is now a leading campaigner for more flexible and compassionate approaches. She has been joined by leading clinicians in the field and the campaign is making progress. For a very moving video account of a recent workshop on the subject see [a].

As one participant commented: 'My doctor said that he wouldn't prescribe an experimental drug because it may have risks, and my Hippocratic Oath tells me I must "first do no harm". If I am willing to take the risk, he should be willing to give me that opportunity.' This is a clash of values that calls out for resolution.

[a] www.youtube.com/watch?v=wilTXvFN2NU

trial centres, and so in fact tend to be young, relatively fit, Caucasian males. Results on females, other ethnic groups, and perhaps most significantly those suffering from other diseases, do not emerge sufficiently from the RCT model.

Patient numbers are those necessary to 'power' the trials to pass the test of statistical significance represented by a p value of less than 0.05: a less than 5% chance that the results could be explained by random variations. However, the largest trial in the sequence, the pivotal Phase 3 trial, cannot contain enough patients to identify the rare serious side-effects that often emerge later, when the drug is widely used.

Also, as pointed out by Mark Walport, the UK's Chief Scientist, the *pressures on regulators are asymmetric*. There is little praise if the product they approve delivers health benefit, but swift criticism if they appear to let a defective product through the net, even if the fault lies elsewhere or was simply unpredictable. The saga of the European breast implants fraudulently manufactured in France with industrial grade silicone, and seemingly inadequately inspected by German plant inspectors, is very instructive. It did not prevent severe media criticism of the UK regulator, the Medicines and Healthcare products Regulatory Agency (MHRA), for allowing them to be used in the UK. It is stories such as this that result in multiple inspections by multiple regulators, with the consequent waste of both public and private sector resources.

It would therefore be surprising if regulatory systems were not risk-averse by nature. For most products, approval allows widespread, and essentially uncontrolled, use by thousands of doctors and millions of patients. Regulators are understandably concerned about rare side-effects and inappropriate use. As one senior regulator commented to me in the context of anti-obesity drugs, developed for morbidly obese patients: 'I have to bear in mind that these drugs may be used by teenagers to get into their prom dress!' This raises the crucial question of the role of the regulator—watchdog or partner in the innovation process? It's interesting to look at the mission statements of regulatory authorities (Box 3.2).

The relative *lack of involvement of patients* in the design of the process—and in determining acceptable benefit/risk—is a hangover from a past, paternalistic medical and regulatory culture. We have already seen how patients are demanding input into trial design. Establishing benefit/risk is next on their agenda. It is surprising how recently patients have been invited into specific benefit/risk discussions. In some cases, such as the multiple sclerosis drug Tysabri, patients had to lobby the FDA to reverse a refusal of a licence on the grounds of a rare but fatal viral disease of the brain that occurred in 1 in 1000 patients. They argued that—provided they were made aware of the risk—the decision to take this beneficial drug should be theirs.

Both the FDA and the European Medicine Agency (EMA) provide scientific advice to companies, but this is typically structured as a series of questions and answers, to avoid the risk that the regulator is coaching some sponsors, perhaps at the expense of competitors. Joint scientific advice by regulators and HTA bodies has the potential to ensure that clinical trials are designed to meet the needs of both. However, uptake of this option is patchy, in part because most companies are structured in a way that separates those responsible for clinical research from those responsible for 'market access' (i.e. reimbursement).

Box 3.2 The mission of the regulator

Is the principal role of the regulator to protect society or to promote innovation? It is somewhat revealing to look at the mission statements of major regulatory bodies:

FDA: 'FDA is responsible for protecting the public health by assuring the safety, efficacy and security of … medicinal products. FDA is also responsible for advancing the public health by helping to speed innovations that make medicines more effective, safer and more affordable. …'

EMA: 'The mission of the European Medicine Agency is to foster scientific excellence in the evaluation and supervision of medicines, for the benefit of the public'

Responsibility for helping speed innovations is a welcome addition to the FDA mission statement and has been reflected both in the various fast tracking, priority review, and breakthrough designation pathways they have introduced in recent times, and also in their enthusiasm for regulatory science, of which more later.

Across the world, the multiplicity of regulators acts as a further burden and brake on innovation. Recognizing this, over a period of 20 years European regulation converged on a single EMA-administered system. The 'first base' was a system of mutual recognition by countries of each others' decisions. This was followed by the current approach for most medicines, in which regulators from EU countries act as 'rapporteurs' (writing the scientific evaluation reports on individual submissions). These are then presented to a cross-country EMA committee called the Committee for Medicinal Products for Human Use (CHMP) for discussion and decision (in the 'centralized procedure'). Even then, the CHMP outcome is a only a recommendation to the European Commission, who make the final approval decision.

Beyond Europe, each country has its own medicines regulator. Many look closely at the FDA and EMA decisions, rather than repeating analyses, but the Japanese and Chinese regulators have fully fledged processes of their own.

Measures and perceptions of risk are obviously central to the process by which new products are allowed on the market. At least two factors complicate this decision in the case of healthcare products: firstly, life and death are often involved and secondly, society and individuals tend to weigh downsides more heavily than upsides.

Of course, life and death regulatory decisions are made in other spheres, like vehicle manufacture, aircraft operation, and fire inspections. As a colleague of mine, Stan Lapidus, pointed out, the US building fire safety regulations are set and monitored by a private sector body, not a government agency, and the results seem none the worse for it! However, there is something very immediate about risks associated with products taken by—or inserted in—the body, and so it is hard to imagine this responsibility being devolved to a non-governmental agency.

Turning to risk perception, social science studies have clearly shown that we are 'wired' to avoid negatives to a greater extent than to seek positives, that is, it would appear, until we are faced with a life-threatening or severely life-limiting illness, as we saw in the case of Tysabri for multiple sclerosis. In my view, the individual patient, once properly informed, is in the best position to make a choice. Assuming the basic efficacy and reasonable safety of a product have been demonstrated, we need to maximize the ability of patients to choose.

It is specific case examples that determine society's and the regulator's appetite for the risk that all new products carry. In Box 3.3, we see three cases that have tested that appetite in recent years and the lessons that we might deduce from them.

Case studies like these reinforce a number of key points in relation to regulation:

◆ Benefit/risk judgments are complicated to make, with data accumulating over time often changing the perspective.

◆ Sincere, professional regulators in different jurisdictions can come to opposing conclusions on the basis of the same data.

◆ The costs to companies of not revealing all the data and analyses they have can be very high.

◆ An ability to identify in advance those patients likely to have side-effects, and exclude them from treatment, could potentially save drug withdrawals.

Box 3.3 Regulatory response to perceived risk associated with new products—three cases

Vioxx

This product from Merck, Sharp & Dohme was one of a new class of anti-inflammatory drugs called cox-2 inhibitors. They were seen as having real advantages over the cox-1 predecessors, as they had much lower levels of gastric side-effects. Hence, the expected safety profile was one of the major advantages, as Vioxx was launched by Merck in 1999, with approval based on eight studies of 5400 patients. The lower gastrointestinal impact of Vioxx versus Naproxen (an earlier non-steroidal anti-inflammatory drug) was clear from these studies; so too was a potential concern about higher cardiac side-effects, although it was initially argued by the company that Naproxen might be protecting against heart problems.

The product proved highly efficacious, giving pain relief and increased mobility to many with serious joint disease. Millions of people were quickly given access to the product. But there were drawbacks, most importantly an effect on the heart rhythm related to a phenomenon called QT prolongation. This effect was of minor importance to most patients but became the suspected source of heart attacks in a small minority.

Once the side-effect problem became clearly established, the company itself decided to take the product off the market, overnight, on September 30, 2004—4 years after launch. Even the regulators, who had been monitoring the situation, were taken by surprise. The company's lawyers felt that to continue marketing the drug would lay the company open to law suits. And so it proved, on a scale that ultimately cost the company around $6bn, including a $950m fine for illegal promotion. And the company's stock price nearly halved in the days after the recall.

Amidst the widespread accusations and recriminations that followed, the real lesson of the Vioxx story was obscured. The product was really effective, and many of those from whom it was withdrawn complained bitterly. The 'risk science' learning was this: if a drug is to be given to a very large population of people without a life-threatening disease, we should proceed with caution, identifying and protecting at-risk populations (in this case anyone with a cardiac complaint) and with careful monitoring. Immediate widespread promotion and use carries a risk for everyone: for patients, clinicians, the company, and the regulators.

Avandia/Rosiglitazone

The glitazones represent a valuable class of anti-diabetic medicines that work by binding to specific fat cell receptors and making them more responsive to insulin. GlaxoSmithKline's product in this class, Avandia, was launched in 1999 and soon became one of the company's best-selling products.

However, reports of cardiac side-effects, particularly in patients with heart failure, began to accumulate, and warnings of the problem were issued in 2002. A series of

(continued)

Box 3.3 Continued

meta-analyses (combing results from several trials) in 2005–2007 confirmed a hazard ratio (increased risk) of 1.3—1.4. Regulators on the two sides of the Atlantic watched carefully but came to somewhat different conclusions at different times. First, an FDA advisory committee voted in 2007 to keep it on the market on overall benefit/risk grounds, then the agency issued black box warnings and then, in 2013, removed restrictions, stating that their analysis now showed no increased side-effects. But the patent on the drug had expired 2 years earlier, and sales were by then at a very low level.

In contrast, the European regulator EMA withdrew the drug from sale in Europe in 2010. GSK was fined $3bn for withholding some of the cardiac side-effect data. As with Vioxx, a number of class action suits ensued, with a possible total cost to the company of $6bn.

Acomplia/Rimonabant

Obesity is one of the most important targets for drug research. It is increasingly prevalent, leads to dangerous levels of diabetes and other complications, and alternatives to drug therapy, such as stomach surgery, are rather costly and interventional.

However, its prevalence means that any effective drug could be given to a very large proportion of the population, whether on- or off-label. Hence safety of products in this class is paramount.

Sanofi's anti-obesity drug, Acomplia (rimonabant), worked as a CB1 receptor blocker, reducing the appetite through effects in the central nervous system (CNS). On launch in Europe in 2006, it was widely expected to be a blockbuster. However, side-effects of its CNS action in some patients were severe depression and suicidal thoughts and regulators saw a worrying level of these reported within months of its introduction, leading to prompt and permanent withdrawal from the market in 2008. It had never been approved by the FDA, which examined it in 2007 and concluded there was insufficient safety data.

The legal and financial consequences for Sanofi were much lower than in the other examples: it reached a settlement on a suit for misleading investors for $40m in the USA.

Could this have been avoided? Were there classes of patients for which the benefit/risk was still acceptably high? We shall probably never know. Could we have tested the mental state of patients and allowed clinicians to use their judgment to give the drug to those with stable mental states whose extreme obesity posed a major threat to their lives?

Conservatism in the regulatory process is beginning to break down, as we will see later. The greater problem now is conservatism on the part of industry. It is not hard to see why. The clinical trial process is associated with huge investments, and the regulatory affairs staff in companies will be roundly criticized, or worse, if they take risks with that investment and their regulatory submission dossiers fail to 'tick all the boxes.' Companies may not like the existing system, but at least they understand it, and have learnt to comply to the letter.

Value and health technology assessment

The second element of the 'T2 gap' relates to *value and reimbursement*. Formerly, once an innovation had surmounted the regulatory hurdles, its acceptance by the health system was simply a matter of convincing clinicians to prescribe, order, or use it. Promotional activities, word of mouth recommendations, and simply growing familiarity with the new product led to its inclusion in treatment.

No longer. The pressure on health systems around the world has spawned a variety of mechanisms to assess cost-effectiveness. This is not only an additional hurdle, it is also one grafted on to the end of the process designed for other purposes. Some of these 'value' mechanisms attempt to be quite precise and scientific, for example calculating the cost per quality-adjusted life year (QALY), as is done at the UK agency NICE (National Institute for Health and Care Excellence). Others seek to assess the level of comparative effectiveness versus existing treatments, or score them on level of innovativeness. Most look for RCTs performed against comparators; other are more pragmatic, recognizing that the eventual comparators in practice are not always available at the time that clinical trials are designed.

Despite all the challenges in gaining regulatory approval, some consider that reimbursement is becoming the greater hurdle, now that many company pipelines are refilling with promising drugs. New products at high prices have become the greatest target for cost control by health systems, and they have put in place increasingly rigorous HTA systems to assist in the process.

These systems are as diverse as the cultures that have spawned them. The UK has perhaps invested the most in HTA methodology to establish value, through the development of NICE (see Box 3.4). The attempt to establish a defensible system results in a rigorous arithmetical calculation of cost per QALY. This is intended to minimize legal challenges. In fact, the only basis for legal challenge of a NICE conclusion is that the advertised process was not rigorously followed, not that it gave an unreasonable answer. There have been examples of the latter, however: the initial NICE conclusion was that someone needed to go blind in one eye from macular degeneration before it was cost-justified to use Lucentis to save the sight in the other eye! The QALY for an innovation is typically assessed using the EQ5D measure of health impact developed by the EuroQol organization (despite its name, it contains several non-European countries, including the USA). It is a simple, self-completed questionnaire of health status.

But what is a year of life worth? This rather basic question has, of course, no simple answer. While most studies in the health arena assume a value of about $50 000 per year, Stanford economists have calculated a value of $129 000 per year, based on retrospective analysis of half a million patients on kidney dialysis[8] Indeed, different arms of the same government (in this case the US) effectively use different values of a human life for different purposes: air pollution ($9.1 m); avoiding smoking ($7.9 m); road traffic accidents ($6m)[9].

Box 3.4 NICE—an experiment in the science of value

In the year 2000, the UK government realized it had a problem: an increasing number of new health interventions, especially drugs, whose cost-effectiveness was unclear. They also recognized that there would be a demand from patients and clinicians to fund these treatments if they were—or appeared to be—superior to what was already available. Politicians did not want to make the potentially very unpopular 'no' decisions themselves: they needed a process and a body to administer it, setting these decisions at arms length from government. The process should be as scientific and objective as possible, able to resist legal challenge when its decisions were particularly unpopular. NICE was born.

Early on, NICE created its own internal mechanisms for independence, setting up appraisal committees, containing clinical, statistical, and economic experts, charged with making the decisions—without reference to NICE management. And they were given access to university-based economic analysis units to construct economic models that compared the costs and health benefits of a new treatment with the existing best practice.

These models calculated the so-called 'ICER'—the incremental cost-effectiveness ratio—in units of pounds sterling per QALY. Cost per QALY of under £20 000 and acceptance for NHS funding was assured; cost/QALY above £30 000 and the technology was likely to be refused reimbursement. In the grey zone of £20 000–30 000, the committee would adjudicate.

The rest of the world watched the experiment with interest. Some other countries—notably Sweden—had already set up their own HTA bodies. Others, like France, had confidential negotiations between companies and government bodies to establish a price to reflect how valuable a treatment appeared. But NICE represented an attempt to make the HTA process as scientific as possible.

Of course, the dominant factor in calculating the ICER was the price that the company intended to charge. In the early days of NICE, this price was a fixed input to the process, with companies very reluctant to reduce it from the levels they were charging internationally. The UK price was (and still is) a major factor in international price-referencing schemes and so concessions in price in the UK would filter down to other countries, reducing profit margins in many other markets.

To avoid such public price reductions, the concept of 'risk-sharing' soon emerged, with companies conceding there were situations in which the price would not apply. One trailblazer discussion was with Janssen over the product Velcade, for multiple myeloma. The response of patients to the drug could be quite quickly gauged from the level of a particular protein in the blood and the company agreed that—if the biomarker did not indicate a response—then the company would not expect to be paid.

Other 'risk-sharing' schemes were devised, such as providing the first few doses or those above some 'typical' number of doses free of charge. Over time, these schemes were less and less popular with the NHS, since they required tracking of

Box 3.4 Continued

a kind that the system was not set up to perform. So increasingly companies were asked to provide confidential across-the-board discounts instead.

Fifteen years on, it is fair to ask: how well has it worked? Predictably, there has often been major controversy, when NICE appraisal committees have denied applications for drugs shown to extend life, especially of cancer sufferers. As a result NICE has introduced 'end of life' criteria to allow appraisal committees to accept ICERs above £30 000 when a treatment offers significant extension of life.

The other major adaptation has been the search on the part of the committees for subgroups of patients where the benefit was particularly strong, and to allow reimbursement for only these patients. This has led to exchanges with industry in which companies believing their product had broader value complained about denial, while NICE regarded their decisions to restrict a treatment to subgroups as a positive—'yes, but' outcome.

NICE's methodology has been widely discussed and copied or adapted in a number of other countries. Even where the NICE approach has not been adopted wholesale, the cost/QALY methodology has spread quite widely. Interestingly, it has been established by IMS[a] that those countries using this metric tend to take up innovative cancer treatments to a lesser extent, resulting in calls for value elements 'beyond the QALY' to be investigated and incorporated in the process.

Quite recently, NICE has taken on responsibility for cost-effectiveness decisions in social care and this has confronted the body with the need to create consistency between the metrics used in health versus social care decisions. Concepts like the WELBY (well-being-adjusted life year) are being investigated. This might lead to elements of well-being that go beyond classic clinical measures emerging from clinical trials becoming part of the appraisal.

Other value elements that are clearly relevant and advocated by those who think NICE's approach too narrow are the value of a return to employment (on the part of either the patient or a carer), the value of a first treatment for a disease for which no therapy exists, and the level of innovativeness that a new treatment represents (especially where this appears to open up a whole new area of potential innovation). Others speak of the 'value of hope', which is obviously most relevant when nothing else exists.

[a] Impact of cost per QALY reimbursement criteria on access to cancer drugs December 2014 (downloadable from www.theimsinstitute.org)

Attitudes to this growing trend of HTA on the part of industry have shifted over time, from outright opposition to reluctant engagement to adaptation to the new demands—but still on a piecemeal basis, country by country. Both HTA agencies and companies have built up large workforces of health economists, analysts, and model builders. They meet annually at the sizeable ISPOR (International Society for Pharmacoeconomics and Outcomes Research) conferences: it is an industry that did not exist a decade or two ago.

The current HTA system is highly fragmented, even where it first emerged in Europe: every member state has essentially its own agency for determining value and/or negotiating prices. This mirrors the derogation of powers over commercial matters to the member states, but the diversity of data needs and evaluation process has become a major burden for companies. Through a body called EUNetHA, there is increasing exchange between HTA agencies on methodology. It makes no sense for the data needs required by each country for their value analysis to vary: the task of establishing comparative effectiveness (the new treatment's value versus current standard of care) is essentially identical. One can foresee a process of either mutual recognition of the results of comparative effectiveness analysis or of a centralized procedure akin to that of the EMA.

Meanwhile the innovator is faced with a patchwork of agencies and methodologies that is extremely difficult for them to navigate, and a set of evidence requirements that are inconsistent and frequently incompatible. It can be 2–3 years before the reimbursement hurdle is surmounted in all major countries, and the product becomes available to patients through the public health system.

Into this confusing landscape, around 10 years ago, came the concept of 'value-based pricing' (VBP). The idea is simple: instead of innovators proposing a price that they believe the market will bear, and then having this tested against individual HTA processes, they should derive their price proposal from the economic value delivered to the health system.

Implementation of the idea is, however, not at all straightforward. Which forms of value to take into account: direct health value, economic and employment value, innovation value, etc? After a few years cogitating about VBP, the UK government effectively abandoned it, in favour of negotiating across-the-board price cuts—simpler for both politicians and industry to understand.

There could still be a resurgence of interest in VBP, but it cannot co-exist long term with international price-referencing, since the value calculated in less affluent markets can then still affect the prices in markets that can and should pay more.

From the innovators' and patients' standpoint, the net effect of the European HTA processes is to slow the introduction of innovations versus, for example, the USA. In the USA there is no such 'cost-effectiveness' assessment (is the product worth its price?), although there is a growing pressure for 'comparative effectiveness' (is this technology better than the one it replaces?). Nevertheless, the US market is not price-insensitive: the major insurance companies aggressively negotiate discounts and the US government benefits from most-favoured nation pricing clauses.

Even so, list prices—and often actual prices—in the USA are often higher than those prevailing in Europe, leading to US complaints that they are bearing the greatest cost burden to provide the innovators with their returns. Likewise, the price disparities between the USA and neighbouring Canada lead to attempts by US consumers (who frequently bear all or much of the costs of their prescriptions themselves) to purchase drugs across the border.

The Obama administration has begun to challenge the current legal prohibition for Medicare to negotiate prices. This will, of course, face strong opposition from industry.

There is a second economic question new technologies face as they enter today's health system. They may be cost-effective, with their cost/QALY below a set threshold, but are they affordable? Complaints about the high price of medicines came to a head in 2015 with the introduction of a medicine, Sovaldi, for the treatment (and cure in most patients) of the chronic disease hepatitis C (HCV). Despite its ability to avoid the huge long-term costs of liver disease (transplants and often liver cancer), at $83 000 per course of treatment, insurers in the USA felt the drug was unaffordable, given the high prevalence of the disease in some parts of the population. Recall that many US patients switch health plans frequently, so avoidance of long-term costs does not weigh as heavily as it might.

In Europe, where most systems are in a position to recognize the long-term returns, there were (and still are) problems. List prices were somewhat lower, but the cost burden in countries with high HCV prevalence was still seen as unaffordable. Negotiations have been tough. The irony is that the drug, when assessed by most of the cost-effectiveness methodologies, emerged as positive, because of the long-term con- sequences avoided. Nevertheless, complained governments and health systems, the bill is unaffordable, and many restrict the drug's use to only the most seriously affected (which of course means that more patients reach this situation than need to).

In the USA, events quickly took a different turn, as a result of two competitive prod- ucts following closely behind Sovaldi. This enabled insurers to strike deals with them, at lower (but still quite high) prices for exclusive supply of these products. This looks like the beginning of a new and more challenging era of price negotiation for innova- tive drugs, which could extend to other therapeutic areas.

In developing countries, Gilead (the drug's manufacturer) asked for dramatically lower prices: the list price in India is 1% that in the USA. This is an important step towards what many believe should be a basic pricing principle in medicines: that prices should reflect GDP per capita.

These examples serve to illustrate that regulatory approval is no longer the most important hurdle for many innovations to surmount. Reimbursement approval is. With a large number of drugs in the pipeline for cancer, in particular, most of which are expected to command high prices, the most significant 'regulator' of whether most patients receive them will probably be these 'value' assessments.

The challenges of value and of affordability will only get worse. The economic pres- sure on health systems will mount as demographics and waves of chronic disease increase the underlying demand for healthcare of all kinds. And we will see the emer- gence of more advanced therapies, such as gene or cell therapy and tissue engineering. These are likely to have a 'one-shot' curative application, with innovators expecting high prices for the single-shot treatments (which will also be much more expensive to 'manufacture' and deliver than most pharmaceuticals).

Innovation in these promising areas will inevitably be held back unless we move to more convergent assessment processes and more creative pricing and financing arrangements (as we will explore in Chapter 5).

The value challenge for diagnostics takes a somewhat different form. The large majority of diagnostics are available at very low prices, with clinicians usually ordering them (typically in the form of blood tests) without much consideration for their cost.

New generations of genomic tests will, however, be one or two orders of magnitude more expensive and are already meeting considerable resistance in publically funded systems. This is despite the fact that, by targeting therapies more precisely, they are able to save many times their cost in therapies that are then not given to those who will not benefit.

A collaboration gap?

All the demands on companies to demonstrate efficacy, acceptable safety, and value have driven huge increases in development costs, as we have seen. It is therefore surprising that they have been *slow to collaborate* on elements of development that offer synergies. Logically, companies should compete primarily on the quality of their products, rather than their ability to assemble a network of investigators, set up a two-arm trial with its own set of controls, qualify, and visit trial sites, and maintain registries of patients with company-specific information infrastructure. Duplication has been rife.

Finally, as will be obvious by now, responsibilities along the innovation process is fundamentally fragmented between agencies. Innovations make their way upstream and have to navigate multiple challenges set up by separate agencies for approving trial designs, organizing trial networks, securing research and ethics approvals, assessing benefit/risk, and adjudicating value. And, apart from the role of the EMA in Europe, these need to be negotiated country-by-country. No wonder time-scales are long and costs high.

There has been little or no coordination between agencies, within and between countries. These bodies ought to see themselves as partners together in bringing products with patient benefit and health system value through the process as rapidly and responsibly as possible, in as globally consistent a manner as possible.

Finally, the process as it stands today has no feedback loop from the health systems that are customers for innovation and the R&D process that produces it. The areas that industry focuses on are those that appear to them to have a reasonable probability of success and support attractive markets. The match with population health needs is quite poor: cardiovascular conditions remain among the top two or three causes of mortality in most countries, and the prevalence of mental diseases is extremely high, but neither represent major targets for current industry efforts.

The challenges for pharmaceuticals at the T2 stage are clearly many and varied (Figure 3.4). Case examples in Chapter 4 demonstrate how they are sometimes allowed to stifle innovation, yet sometimes ways are found to overcome them. When it comes to solutions (in Chapter 5), though, it is also clear there will be no 'silver bullet' to address all these issues.

T1 and T2 in medical devices

Regulation and reimbursement of medical devices have received less attention than those for pharmaceuticals, but there are still significant issues. As a review of the field commented: 'Despite the rapid advancement of state of the art medically driven technologies, numerous challenges still exist for translating technology to the clinic including long and often undefined regulatory approval pathways, high translational

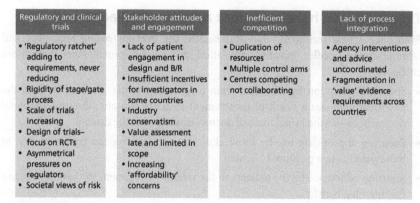

Regulatory and clinical trials	Stakeholder attitudes and engagement	Inefficient competition	Lack of process integration
• 'Regulatory ratchet' adding to requirements, never reducing • Rigidity of stage/gate process • Scale of trials increasing • Design of trials– focus on RCTs • Asymmetrical pressures on regulators • Societal views of risk	• Lack of patient engagement in design and B/R • Insufficient incentives for investigators in some countries • Industry conservatism • Value assessment late and limited in scope • Increasing 'affordability' concerns	• Duplication of resources • Multiple control arms • Centres competing not collaborating	• Agency interventions and advice uncoordinated • Fragmentation in 'value' evidence requirements across countries

Fig. 3.4 T2 issues.

costs, potential safety concerns with nanomaterials, assessing appropriate risk/benefit and cost/benefit ratios, and appropriate matching of technology and application.'[10]

This is a US commentary. There is a major difference between the regulatory systems in the USA and Europe for medical devices. While the FDA centralizes the approval process (via a PMA—pre-market approval), the European process is decentralized. Individual 'notified bodies' issue 'CE marks', based on clinical trials, and these are recognized across the EU.

The net effect of this difference is a comparative delay for many devices entering the US market, where the regulator is criticized for making 'the perfect the enemy of the good'[11]. Of course, there are issues in medical device manufacture, from both the hardware and software perspective, which do not exist for pharmaceuticals, and so it could be argued by regulators that rigorous overview of this aspect of device approval is vital.

Examples of these comparative delays are as follows: a novel left ventricular assist device for heart failure patients—29 months; a pacemaker to resynchronize heart contraction—30 months; a minimally invasive transcatheter heart valve for inoperative aortic stenosis—more than 4 years. The concern in the USA is that the faster process in Europe drains the US of some of the leading edge clinical research capacity that it has long enjoyed, as the leading nation innovating medical devices.

The CE mark process that seems to deliver faster market approval in the EU has had relatively few problems, only the defective breast implants mentioned elsewhere and a surgical sealant have had to be removed from the market for safety reasons.

Reimbursement for high-cost devices is probably the reverse situation: the high-cost device, once approved, becomes standard of care more rapidly in the USA than in Europe, where payers struggle to see the long-term versus short-term economic consequences of approving them. We will examine the example of implantable defibrillators in Chapter 4.

Once the multi-step, multi-year, and very costly process leading to product approval and reimbursement has been completed, one might expect that most of the problems would be behind an innovation. Sadly not.

T3: why do we have such poor adoption and adherence?

Even once a new technology has surmounted the ever-increasing regulatory and reimbursement hurdles, effective translation into patient benefit faces three more challenges:

◆ Spreading adoption by clinicians from a few centres that are familiar with the new treatment—as a result of involvement in trials and attending international conferences—to the broad mass of doctors dealing with the disease

◆ Ensuring appropriate use by those clinicians, prescribing the treatment just to those patients in a position to benefit

◆ Securing adherence by the patients to the treatment regimen, so that they receive the intended benefit

Clinical adoption and appropriate use

This is often left to the innovator company's promotional efforts, with the health systems standing by, advising doctors to restrict use, or even 'counter-detailing', if they feel the new treatment is more costly than they can afford.

So uptake by the broad clinical population can take many years, with the median time from product launch to maximum usage for medicines averaging 8.2 years[12].

Many physicians wait to learn about the experience of their more adventurous peers before adopting themselves. When asked what barriers innovation uptake encounters in their practices, UK general practitioners quote a large number (Figure 3.5).

The familiar Rogers curve depicting the phases of adoption of a new technology also applies in medicine[13], with the laggards sometimes a decade behind the early adopters.

Inappropriate prescribing results from a mixture of poor clinical education and limited knowledge of 'stratified medicine'—the use of biomarkers to define and target patient populations for whom the benefit/risk ratio will be highest. So the proportion of patients benefiting from treatments commonly varies in the range of 40–60%, representing an enormous waste of resources and/or risk to patient health.

Patient adherence

Between a quarter and two-thirds of patients fail to comply with therapy for long-term conditions. A 2008 study compared adherence rates in the first year of prescription of seven major conditions: 72.3% of hypertension patients achieved adherence rates of 80% or more, compared with 68.4%, 65.4%, 60.8%, 54.6%, 51.2%, or 36.8%, respectively, for those with hypothyroidism, type 2 diabetes, seizure disorders, hypercholesterolemia, osteoporosis, or gout[14].

We are beginning to understand some of the underlying reasons for non-compliance. My colleague Rob Horne has pioneered a necessity-concern model, in which patient attitudes to whether they feel the prescribed therapy is important to them (necessity) and whether they are worried about downside effects such as drug dependence or side-effects (concern)[15]. Analysis of the adherence literature across more than 90 papers shows, not surprisingly, that high adherence is associated with stronger perceptions of the necessity of treatment[16]. Adherence is particularly low with anti-depressants (19% in one study) and concern about side-effects is a major driver of non-adherence[17].

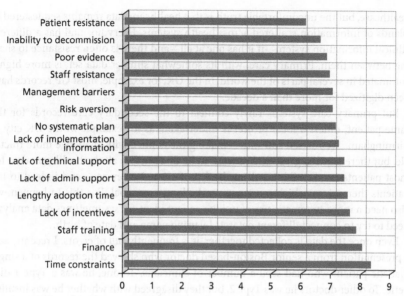

Fig. 3.5 Perceived barriers to innovation uptake.
Reproduced with permission from Nesta. *Which Doctors Take Up Promising Ideas?: New insights from open data*, http://www.nesta.org.uk/sites/default/files/which_doctors_take_up_promising.pdf, accessed 01 Oct. 2015. Copyright © 2014 Nesta.

All of this analysis indicates a weak partnership between the patient and the physician: adherence requires the latter to have identified and dealt with necessity-concern issues on the part of the former. In fact, the terminology here is revealing: doctors often speak of 'compliance', or 'adherence'—both implying a rather passive stance on the part of the patient. 'Do this—and trust me you'll feel better', as opposed to active engagement in treatment choice and application.

When we learn that patients have failed to fill prescriptions, have stopped taking therapies after a few weeks, or have taken 'drug holidays', the system tends to assume they have simply forgotten and need to be reminded.

T4: are we learning from real world use?

Definitely not. Here are some of the questions to which we desperately need answers, none of which are available on any systematic basis:

- What are the real world outcomes of therapies in normal populations, and how do they compare with the clinical trials?
- What off-target effects are we seeing, either beneficial or detrimental?
- How are multiple therapies working, or interacting, in patients with several diseases?

While information and its analysis is leaping ahead in other sectors, the healthcare sector lags by perhaps 20–30 years. There are many conferences on Big Data in healthcare, but the uncomfortable truth is that healthcare data remains sequestered in islands of information scattered across health systems. Every hospital has a different clinical information system—if it has one at all—and there is often resistance to sharing between them. Primary care, with its somewhat simpler data set, is more highly automated in several parts of the world: in the UK, for example, most GP records have been digitized for more than a decade.

But primary care systems rarely connect to the secondary care records for the same patient, unless a local project is undertaken, as for example in the UK city of Birmingham. Web-based software is making the connectivity challenge more tractable, but there is still a great deal of work to be done before all the healthcare data for most patients is accessible to all medical professionals they see—and indeed to the patients themselves, which must remain the long-term goal. For this to happen, we also need a more systematic use of unique patient identifiers: sometimes data analysts need to try to connect different datasets by using name and date of birth.

Even once the data is collected together, it is frequently full of errors. I recently saw a presentation from a senior Boston-based doctor who showed the records of a single diabetic patient who had seen a number of clinicians. To one, he was a Type 1 diabetic. To other doctors he was Type 2, but they disagreed with whether he was insulin-dependent or not! So we need incentives or sanctions to ensure the data entered into clinical systems is accurate and comprehensive.

Without a functioning, accurate, and connected IT infrastructure delivering reliable outcome data, it is hard for the system to learn from the real world use of innovations. This is not just important in their proper evaluation, it would also give important insights into the development of future products, by identifying the characteristics of non-responding patients.

The most intriguing output from Big Health Data analysis is the identification of products prescribed for disease A that seem to have protective or palliative effects on disease B in those patients suffering from—or at risk for—both. As we pointed out earlier, this is exactly the kind of information we cannot derive from RCTs on patients carefully selected as sufferers of only one condition.

The UK has made several forays into assembling health data for epidemiogical or other research purposes. The Clinical Practice Research Datalink contains around 10 million primary care data records. The Health & Social Care Information Centre has high level information on the whole NHS population. In the USA, major systems such as Kaiser Permanente and Geisinger have very rich datasets on their patients and treatments. All these data are mined by companies to assess the effectiveness of their products. So all is not gloomy, but there is still a great deal of ground to make up if the healthcare sector is to catch up with most other information-rich sectors. As a result, despite much talk of evidence-based medicine, the data is thin and very dependent on specific trials.

In order to secure real feedback on the effectiveness of medicines, it is often necessary to develop product-specific patient registries, and companies frequently do this when asked by regulatory authorities to follow up on outcomes. But they see this as a costly endeavour.

The different challenges of diagnostics and devices

The path to patient benefit for drugs, as we have seen, is long and complex, with multiple gaps to bridge. The story in devices and diagnostics has parallels but some of the challenges are quite different, as we will see in this section.

Diagnostics

The in-vitro diagnostics industry is the Cinderella of medical innovation. Few, if any, important medical decisions are made without their use, yet they attract much lower levels of pricing and therefore investment than the therapeutics that are prescribed on the basis of the diagnostic results.

With these poorer potential returns on investment the cost of rigorous clinical validation is hard to justify. However, once there is evidence of reliable performance—typically measured by sensitivity and specificity against the previous 'gold standard' of diagnosis—regulation is less burdensome than for therapeutics, usually being covered by the FDA 510k approach (a comparison of the new product with the existing standard product).

In Europe, diagnostics do not go through the EMA, the FDA equivalent, but a series of national 'notified bodies'. So, in contrast to the situation in pharmaceuticals, Europe appears a more attractive rapid launch market for some diagnostics. The latest revision of diagnostics regulations in Europe, however, has tightened scrutiny over those diagnostics that carry significant risks, such as those that identify viruses or cancer.

The greater challenge for modern diagnostics, including those based on gene identification or protein quantification, is reimbursement. In the USA, they can often be assessed and reimbursed on the basis of the technical steps involved in carrying out the test (the so-called 'stacked codes' approach). In Europe, however, they are often purchased by individual laboratory or hospital providers to whom their hundreds-of-euros price tags seem exorbitant against the few euros per test for conventional diagnostics.

As a result, even through the test may be directing therapies that cost thousands of euros, the diagnostic budget-holders may be reluctant to accept them, resulting in an overall mal-optimized situation in which high drug costs are wasted while low diagnostic costs are well controlled.

Devices

Devices, of course, come in many forms: electrical cardiac support, neurological brain stimulators and pain supressors, hearing aids, and limb and joint prostheses.

As we have seen, the regulatory regime differs on the two sides of the Atlantic: centralized FDA approval in the USA, national notified body and CE marking in Europe. The tougher environment in the USA has encouraged many manufacturers to trial and register their products in Europe first, the reverse of the normal pattern for drugs. A 2010 survey by Advamed, the US trade association, demonstrated that the large majority of companies found the EU-based authorities more predictable, reasonable, and transparent. The EU recently introduced revised regulatory standards, however, and the relative attractiveness may shift.

Development costs can be lower for devices than for many drugs. Low to moderate risk products go through a so-called 510k process in the USA: this costs around $30 m. Higher risk products going through a new marketing approval (PMA) process cost around $100m to develop. The timescale for the latter (from initial contact with the FDA to approval of a PMA) was 54 months on average, five times the 11 months required in Europe[18].

Reimbursement is a different story. In Europe, as with drugs, each country (and sometimes each payer) sets or negotiates a price for their market. In some cases that is via agreeing the cost of the overall procedure in which the device is implanted. These are typically controlled in a diagnostic related group (DRG) system. This means that a substantially improved and more expensive product has to trigger a new DRG, which can be a lengthy and cumbersome process[19]. Some countries have put in place an HTA process for devices, although there has generally been less experience of this than with pharmaceuticals. In the USA, the FDA has been piloting a joint FDA-CMS (Centers for Medicare and Medicaid Services) review to shorten the time from FDA clearance to reimbursement by the major payers. With such significant discrepancies between systems, costs, and timescales in devices between major markets, the need and opportunity to create 'best global practice' is clear.

The only part of the 'T3 gap' to worry about for most devices is spreading knowledge and skill across the clinical community. Here manufacturers have a stronger relationship with many surgeons than the pharmaceutical industry does with prescribers, with their personnel often accompanying them into surgery. Patients, once the device is implanted, have relatively little option but to comply!

Adoption of many devices remains a slow process, especially where clinician training is a barrier. In the UK, the NICE guidelines for the use of insulin pumps for poorly controlled Type 1 diabetics anticipates that around 15% of patients will need them. This technology does not just improve hour-to-hour glucose control, but reduces the number of hypoglycaemic events that result in costly emergency admissions, or worse. In many parts of the NHS, the proportion is less than 5%. The principal problem seems to be clinical familiarity with the pumps themselves. However, despite positive NICE guidance, some payers may be refusing reimbursement of the high up-front cost because they are focused on short-term, not longer-term, economics.

We have now reviewed the T1, T2, and T3 gaps for the major classes of product. Many of the problems we have found are about mechanisms: mechanisms for regulatory approval, procedures for reimbursement, and a lack of mechanisms for spreading new technology and encouraging patient adherence. However, there is also a cultural backdrop which inhibits collaboration between the three main stakeholders—academic researchers, commercial companies, and practicing clinicians.

The underlying cultural issues

The medical innovation process is blighted by an unusual number of 'culture clashes' among those who need to collaborate to transform its productivity. These affect their

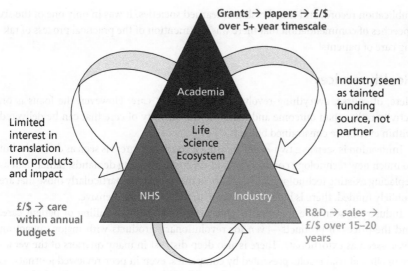

Fig. 3.6 Significant cultural gaps.

perceptions of priorities and of value, their approach to risk, and therefore what they ask of the 'bargain' between innovators and the society whose needs they are seeking to address. The different drivers and their impact on the relationships between the stakeholders are depicted in Figure 3.6.

Let us start by looking at the three main protagonists and how they think about themselves and the others.

Academics

Academics are broadly aware of their seminal role in the life science enterprise but have relatively little understanding of the activities downstream of their discoveries. Their principal concern is securing the grants that keep their groups running and their personal stature in the scientific community. The latter—and often the economics of the departments in which they work—is driven by their publications, both the quantity and quality, the latter assessed by the 'impact factor' of the journals in which they publish. Priority in publishing is all, and this has often led to the publication of irreproducible results.

Some academics are entrepreneurial and get involved in spin-out companies and participate in advisory boards of larger life science companies. However, the prevailing attitude to the R&D done in the commercial sector is that it is not as scientifically prestigious as their own and that there is something not quite respectable about the whole industrial sector, with its focus on profit.

Many academic clinicians seem to regard the routine care of patients as less interesting and valuable than research. I once attended the honorary fellowship ceremony of a UK clinical body, in which top doctors lauded each other for their prize lectures, publication record, and membership of learned societies. It was in only one of the six speeches of commendation that there was any mention of the practical process of taking care of patients!

Health services

Here, of course, everything revolves around patient care. However, the focus is on activity rather than outcome and therefore the quantity of care that can be delivered within ever more constrained budgets.

Innovation is seen as a two-edged sword. Of course, progress is seen as desirable, but so much new technology comes as an add-on cost to existing demands, rather than as replacing existing technology and cost. So, in many systems, particularly those that are publicly funded, there is built-in resistance to the new and expensive.

Industry is seen as a less worthy enterprise than the high calling of patient care, and the prices of products—even for revolutionary products with major impact on lives—seen as extortionate. There is also deep distrust in many quarters of the veracity of clinical trial results presented by companies, even in peer-reviewed journals, as negative results are frequently not reported. Some clinicians do spend time in industry but it is usually a one-way street.

There is often also a cultural gap towards the academic enterprise, although of course in major medical centres many doctors play both roles. There is frequent mention of the 'front line' and the need to protect or advance it when resources need to be allocated, not always recognizing that research is the front line for the future patient.

Industry

The lack of real comprehension of the other players continues in much of industry. Most executives enter the life science industry with a genuine interest in improving the lot of humanity. However, for those working in public companies the consistent pressure of the shareholder community for increasing quarterly earnings inevitably has a major influence on their priorities.

There is recognition that without the basic research of academics the stream of innovation would run dry. However, industry is increasingly sceptical of directly supporting academic researchers as they need to publish quickly (sometimes obviating the patent protection that companies seek) and many of the results cannot be replicated in the sponsor's laboratories.

Within industry there is also often an internal cultural gap—between R&D and marketing and sales. The former value their relationships of intellectual interchange with leading academic researchers, but the latter focus more on recruiting key opinion leaders who are seen as central to the marketing enterprise. Their conference speeches and presentations to colleagues represent a powerful source of endorsement. While there is increasing levels of scrutiny of the financial relationships between companies and key opinion leaders, there is still suspicion among the general medical profession that it is the money rather than the quality of the product that speaks when they present.

So, while all these three groups are critical to the innovation process, they march to very different drummers. Of course, much of this is inevitable, but we need to dispel ill-founded prejudices and forge more productive working relationships. We will return to this in Chapter 8 on the future ecosystem.

Summary of Chapter 3

There are no fewer than five major gaps in translation in the long journey from discovery to practical patient benefit. Insufficient understanding of disease mechanisms (T0), limited skills and motivation in turning lab discoveries into potential products (T1), huge wastage in bringing promising products to market (T2), disappointingly slow adoption by doctors and adherence by patients (T3), and a failure to learn from the outcome of past treatments (T4)—all of these cripple the productivity of life sciences. And major cultural barriers between academia, practical medicine, and industry make matters worse. The net result is poor yield at every stage in the innovation process and therefore very poor translation overall.

In Chapter 4, we look at a wide range of past successes and failures and ask: what lessons can we learn as we seek to redesign the translation process? Can specific successes furnish principles for better translation? Can we understand what has held us back in areas of poor progress, and design our reforms to make a decisive difference?

References

1. Tufts Clinical and Translational Science Institute. http://www.tuftsctsi.org/About-Us/What-is-Translational-Science.aspx, accessed 01 Sep. 2015

2. DeJesus-Hernandez M, Mackenzie IR, Boeve BF. 2011. Expanded GGGGCC hexanucleotide repeat in noncoding region of C9ORF72 causes chromosome 9p-linked FTD and ALS. *Neuron*, 72:245–56

3. Walsh J, Arora A, Cohen W. 2003. Research tool patenting and licensing and biomedical innovation. Paper presented at: Copyright and Database Protection, Patents and Research Tools, and Other Challenges to the Intellectual Property System; November 24–25, 2003; Maastricht, the Netherlands

4. Prinz F, Schlange T, Asadullah K. 2011. Believe it or not: how much can we rely on published data on potential drug targets? *Nat Rev Drug Discov*, 10:328–9.

5. Begley G, Ellis L. 2012. Drug development: raise standards for preclinical cancer research. *Nature*, 483:531–3.

6. Barker RW, Scannell JW. 2015. The life sciences translational challenge: The European perspective. *Ther Innov Reg Sci*, 49:415–24.

7. Sir Michael Rawlins, 2008, Harverian Oration, Royal College of Physicians.

8. Staff Paper, Center for Social Innovation, Stanford Graduate School of Business, http://csi.gsb.stanford.edu/how-much-will-we-pay-for-year-of-life, accessed 01 Sep. 2015

9. Binyamin Appelbaum, 'As US agencies put more value on a life, Business Fret' *NY Times* Article dated 16 Feb. 2011 http://www.nytimes.com/2011/02/17/business/economy/17regulation.html?_r=0, accessed 01 Sep. 2015

10. Karp JM, Langer R. 2007. Development and therapeutic applications of advanced biomaterials. *Curr Opin Biotechnol*, 18:454–9.

11. Citron P. 2011. Medical devices: lost in regulation. *Issues Sci Technol*, 27(3).

12. Dunn AG, Braithwaite J, Gallego B, Day RO, Runciman W, Coiera E. 2012. Nation-scale adoption of new medicines by doctors: an application of the Bass diffusion model. *BMC Health Serv Res*, 12:248.

13. Everett Rogers, The diffusion of innovation, http://blogs.bmj.com/bjsm/files/2013/01/the-diffusion-of-innovation.jpg, accessed 01 Sep. 2015

14. Briesacher BA, Andrade SE, Fouayzi H, Chan KA. 2008. Comparison of drug adherence rates among patients with seven different medical conditions. *Pharmacotherapy*, 28:437–43.

15. Horne R, Weinman J. 1999. Patients' beliefs about prescribed medicines and their role in adherence to treatment in chronic physical illness. *J Psychosom Res*, 47:555–67.

16. Horne R, Chapman SC, Parham R, Freemantle N, Forbes A, Cooper V. 2013. Understanding patients' adherence-related beliefs about medicines prescribed for long-term conditions: A meta-analytic review of the necessity-concerns framework. *PLoS One*, 8:e80633.

17. Hunot, V, Horne R, Leese MN, Churchill RC. 2007. A cohort study of adherence to anti-depressants in primary care: the influence of antidepressant concerns and treatment preferences. *J Clin Psychiatry*, 9:91–9.

18. FDA Impact on U.S. Medical Technology Innovation: A Survey of Over 200 Medical Technology Companies ◆ November 2010. http://advamed.org/res.download/30, accessed 01 Sep. 2015

19. Mattias Kylstedt, How medical devices are reimbursed in Europe http://www.rspor.ru/mods/congress/stokg13/10.pdf, accessed 01 Sep. 2015

Chapter 4

Lessons from the last 20 years

Despite the formidable barriers to innovations described in Chapter 3, all is not doom and gloom. We have seen many successes that buck the trends and overcome the obstacles. In the last two to three decades we have seen effective therapies for the once-dreaded HIV, fast development of pandemic vaccines, effective eradication therapy for duodenal ulcers, and a new era of precisely targeted medicines for cancer and cystic fibrosis. Even crises like Ebola have brought out some of the best in the R&D enterprise.

However, there are also lessons to be gleaned from our disappointments. We have struggled to come to terms with antimicrobial resistance (AMR), neurodegenerative diseases like Alzheimer's and some of the most crippling diseases of children and adults, like ALS and muscular dystrophy, and we can draw out pointers to better future strategies.

Some of the stories that follow will be familiar to some of my readers. But their purpose is both to bring alive the problems described in Chapter 3 and also point towards the possible solutions—those we will lay out in Chapter 5.

HIV therapy

From what once was a mystery disease with an inevitably and rapid fatal outcome, HIV—if properly treated—is now a chronic disease that no longer shortens lives. How did this come about? And why? And what are the lessons for other, yet unconquered diseases?

In 1981, the emergence, initially in San Francisco, of a strange and fatal infection that crippled the immune system puzzled local clinicians and then began to appear in many cities and countries across the world. The identification of a novel virus, by Gallo and Montagnier in 1983, began a search for anti-virals able to tackle what we now know to be a very elusive retrovirus. We now know, of course, that the virus had been circulating in African primates for many years and had passed (via bushmeat) to humans on at least three occasions, possibly as early as 1910, with the first documented case in 1959.

The challenge of the virus is its ability to enter several of the cells of the immune system (meant to attack invaders) and harness their own transcription machinery to convert the virus's single-stranded RNA into double-stranded DNA (hence the term retrovirus). This DNA is then incorporated into the host cell's nucleus and can remain there, latent and unrecognized by the immune system. There has been a huge amount of research into the life cycle of HIV and its means of spreading in the body, and this has resulted in a number of possible drug intervention points.

The first effective therapy, AZT (azidothymidine, also known as retrovir), an inhibitor of nucleoside reverse transcriptase (NRT), had been previously studied for its effect on the replication of other viruses. It came from NIH sponsored research, was commercialized by the Wellcome Foundation, and approved by the FDA in 1987. This first NRT inhibitor gave some relief from the onward march of the disease.

The first pharmaceutical specifically designed around the HIV-specific protease (the viral protein required for the budding of new viral particles from host cells) was approved in 1989. In combination with two NRT inhibitors, the protease inhibitors cut death rates to a quarter of those seen previously. Ten such inhibitors have now been developed.

We then saw a succession of other drugs that intervene at different stages of the virus life cycle, such as fusion inhibitors (preventing HIV entry into the host cells), variants on the NRT inhibitors that block the action of the reverse transcriptase enzyme in a different way, and integrase inhibitors, that prevent the integration of the viral DNA into the host DNA.

The virus lacks so-called 'proofreading' enzymes to correct errors that occur in viral replication (a process that takes place in less than 2 days). This enables HIV to mutate rapidly. As a result, resistance to therapy can rapidly develop, especially when the initially complex drug regimens are taken improperly. Hence the need for drug combinations, now incorporated in fixed dose combinations produced by products from more than one company, with good adherence from patients.

This story is not just a triumph for pharmaceutical development, but also for diagnostics. The emergence of CD40 and HIV viral load tests confirmed the presence of infection, but also led to the insight that treatment should be started as soon as possible, before the viral load started to climb.

HIV patients were far from the passive recipients of these rapid scientific, industrial, and regulatory advances: they demanded it. Mac Lumpkin, formerly of the FDA, recalls his first day at the agency, when HIV activists threw blood on to the walls of the agency in an attempt to galvanize action. Sufferers of this disease turned quickly into activists, the more so because they thought society was leaving them to their fate. So they organized. In the USA, the charge was led by ACT UP, formed as early as 1987 at a gay community services event in New York addressed by Larry Kramer, with the challenge: 'Do we want to start a new organization devoted to political action?' This high level of activism has continued ever since, and has been a major force in driving acceleration in the regulatory process.

Combination of second generation HIV anti-virals drugs acting on different targets is known as highly active antiretroviral therapy (HAART). This therapy—now available widely in the developed world—decreases the patient's 'viral load' and effectively turns the infection into a chronic disease. HIV-positive patients benefiting from HAART therapy now usually die of other causes, while still harbouring the virus is various reservoirs in the body[1].

HIV, of course, has long ceased to be a disease of the homosexual community in the USA—it affects heterosexuals and the children of sufferers and has become a truly global pandemic. Despite the efforts of the Global Fund and other philanthropic enterprises, HIV still takes lives in the millions in Africa, and in other geographies that struggle to afford the second generation combination anti-virals that control the disease so effectively in the developed world.

So the battle against HIV is far from over. Different subtypes of the virus continue to appear that mislead diagnostic tests and that may not succumb to the same antivirals. And, even in the commonest subtypes, resistance still emerges. Complete eradication of the virus is now therapeutic researchers' highest priority. The virus lurks in reservoirs around the body, even when circulating viral load is low, so an effective 'cure' will require flushing out these reservoirs.

The most fundamental breakthrough we await is that of an effective vaccine. Since the virus had been identified in the early 1980s, there was an expectation that we would have an effective vaccine in the same decade. However, not only does the virus hide in the cells it infects, it also exhibits great variation in the so-called 'epitopes'—the sections of its cell surface proteins visible to the immune system.

Some vaccines have proceeded as far as Phase 3, but these typically show a reduced transmission rate but not complete protection.

The lessons of the HIV story

Patient power

HIV patients, their friends and families—especially in the USA—were unashamedly vocal in their calls for treatments to be developed, approved, and funded. What started as individual cries for help became a movement, embracing the media and community and religious groups in a way that politicians and regulators could not ignore. *Informed patient activism works.*

Media frenzy

Once the means of transmission and rapid spread of the disease became clear, the HIV problem took on the profile of an epidemic. Even governments abandoned the usual safe, boring, and ineffectual ways of communicating risk and addressed the population in terms of a dark and lurking danger: I recall the dramatic public health ads of the UK government in the 1980s. *Infectious disease disproportionately activates media attention.*

Regulatory flexibility

HIV was the area that essentially created the FDA 'fast-track' designation. Multiple new products sped through the new route with less than the usual evidence of long-term safety. The interesting observation made to me by an ex-FDA official, was that none of these 'fast-tracked' medicines ever needed to be withdrawn. So, in practice, no safety risk was run by the patients receiving them. *Regulatory flexibility can emerge when need is seen as urgent.*

Industry collaboration on combination therapies

It was quickly realized that a single drug was much less effective than double- or triple-therapy. This forced companies to collaborate and regulators to also fast-track these combinations, so that the patients did not need to take the 30 or more pills a day that early HIV patients required. *Companies collaborate when the need to do so is clear.*

Anticipating and overcoming resistance

Many drugs face resistance problems, as biology adapts to counteract the effect of a drug, but viruses do so extremely rapidly, mutating as they divide. This means the whole field—both in terms of drug development and therapy compliance—had to be designed to anticipate and head off the emergence of resistance, both for the sake of the individual patient and the effectiveness of treatment for the whole population. As we see later, drug resistance is not confined to infections: any cell or organism that can mutate can evade drugs. *Drug resistance needs to be anticipated.*

The role of international aid and philanthropy

While the disease was being brought under control for millions of patients in the developed world, millions more in developing nations were dying of AIDS, and this eventually became the spur for a type and level of international health intervention we hadn't seen before. Private sector investment is inevitably geared to economic markets, but—in a globally networked world—not having a solution for those unable to pay the prices of the developed North was increasingly untenable. *We need effective strategies for the developing world, or innovators will always be under attack.*

While of course some of the features of this story are rather specific to HIV/AIDS—particularly the young, outspoken patients and the sense of impending epidemic—most of these lessons can be adapted and adopted for other areas of urgent health need. Giving patients a voice and regulators the freedom to adapt to urgent circumstances are perhaps the most powerful tools we can carry into our overall design of the future innovation process.

But I don't want to leave the impression the story has been wholly positive or that it is now over. The HIV field has seen battles between industry and governments over intellectual property (IP). The catastrophic decision by industry to sue the government of South Africa led by Nelson Mandela for flouting international IP regulations cast a long, negative public relations shadow. Paediatric formulations of drugs were not—and still are not—as readily available as they should be. Creating an effective vaccine and eradicating the virus from HIV-positive patients continue to absorb much leading edge science and clinical development. Much of the HIV story is still to be written and further translation challenges await.

Pandemic flu

Influenza is far from a harmless infection at the best of times. It results in the deaths of tens of thousands, typically the frail and elderly, every winter. However, every few years a strain of flu emerges—typically from either its swine or avian hosts—into the human population with much greater levels of virulence and mortality than usual. It was in 2009 that the swine flu strain H1N1 arrived: influenza strains are termed HnNm to describe what forms of the antigens **H**aemagglutinin and **N**eurominidase are expressed on the surface of the virus. (The 'H' binds the virus to red blood cells, while the 'N' helps the virus bud from infected cells).

H1N1 posed such a global threat that the World Health Organization (WHO) and governments around the world deemed sufficiently serious to term it a 'pandemic'. (The widely accepted definition in the Dictionary of Epidemiology is an epidemic that is 'occurring worldwide … crossing international boundaries and usually affecting a large number of people'. Note that severity is not part of the definition.)

One of the reasons the H1N1 strain of 2009 was regarded as unusually dangerous was that genetic analysis showed that it was much more than a typical swine flu: it was the result of genetic reassortment and contained genetic fragments from two different swine flu strains, one avian flu, and one human flu variant[2]. As the magazine *New Scientist* commented at the time 'an unusually mongrelized mix'.

This mongrel seemed much more dangerous than the usual seasonal flu. Indeed, by the beginning of 2010 H1N1 had caused 17 000 recorded deaths. However, with the wisdom of hindsight, the threat was overstated. But industry, the WHO, governments, and regulatory authorities all acted in the belief that H1N1 was a very dangerous infection, and so we can learn from this event how rules can be rewritten in such situations.

The development of a flu vaccine is typically performed on a well-established annual cycle. Viruses that are already circulating, typically in Asia, are collected. The three or four of those that are seen as most likely to cause illness are selected by a process coordinated by the WHO and the US Centers for Disease Control (CDC). This typically happens in February, and cultivation of vaccine (in eggs or a cell culture system) begins, for use in the following northern hemisphere winter.

Flu infections come in waves a few months apart, so in practice the vaccine is available for the second wave. In the case of H1N1 in 2009, the first wave appeared in both the USA and Mexico through April and May, with the pandemic being declared in June, the first such declaration since 1968.

So vaccine development for H1N1 had to be aimed at the second wave, expected over the winter of 2009/2010. This was a very challenging time-frame: to go from genetic analysis to vaccine manufactured and distributed at scale in half the usual time required both accelerated product development and very rapid regulatory and reimbursement review.

At the time of the H1N1 outbreak, industry was already making batches of vaccine for the following northern winter, based on strains that were circulating prior to H1N1. There needed to be a rapid switch to a new vaccine, and industry declared they were ready to make the switch at a meeting with the WHO in mid-May. The first company to have a vaccine at scale, Novartis, had it available by July, with a number of other companies only days or weeks behind. Urgent testing of potential vaccines began in August, with a 6-week trial conducted by the NIH in the USA.

The EMA, like its counterpart the FDA, was faced with the need to approve a new vaccine with very limited human testing. By October 2009, the agency had already issued a detailed strategy for benefit/risk monitoring, accepting that there would not be time for lengthy conventional human trials[3].

They also built and ran decision analysis models, to determine whether delaying approval by a month for more safety data to be collected was likely to save more lives than approving the vaccine immediately. In fact, under all reasonable assumptions, the EMA's models demonstrated the opposite: any delay would cost more deaths[4]. The assumptions in the model that the analysts used about the virus, its spread, and the likely benefit/risk of the vaccine, based on past comparable experience, could of course be challenged. However, the decision was largely vindicated, in terms of the safety of the vaccines that emerged.

According to the WHO, 65 million doses of vaccine had been administered worldwide by mid-November, with serious adverse events in only 1 in 200 000 doses, less than the normal rate of such events for seasonal flu[5].

The lessons of H1N1

Sense of urgency in the global health community

The threat of millions of deaths galvanized activity and public–private and interagency collaboration on a global basis. Of course, infections are not the only such threat, but they do energize the community powerfully. *A sense of urgency helps stimulate collaboration.*

Strong governmental and public engagement

These laid the foundation for industry and regulators to move at an unprecedented speed. Although there were tough negotiations on pricing, it was also much easier to reach agreement when a deadline was in place and the societal need was (or at any rate seemed at the time) overwhelming. *Top-level governmental action can change the rules or compress normal timescales.*

Adaptive development and regulatory flexibility

The urgency also enabled regulators to allow an accelerated development path, dispense with some of their own routine steps and make reasonable assumptions based on past experience. *Development and approval processes can be adapted.*

Benefit/risk modelling

The use of tools for benefit/risk modelling to determine whether delay for further safety evidence was in the public or patient interest surely has wider applicability. This could be applied to a non-infectious disease that has serious consequences for current sufferers. *Modelling can support the case for change.*

The most obvious lesson, of course, is the potential value of a universal flu vaccine, a challenge on which industry is now working. The challenge is the level of prior immunity in the human population to potential components of a new vaccine: one promising approach is to use a flu virus to which few humans have been exposed, for example, one known as PanAd3. This virus can infect human cells and cause them to produce two antigens (M1 and NP) that are found across most, if not all, flu viruses, but the virus does not replicate itself in humans. Early animal trials at the FDA show it confers immunity to subsequent flu infection and also triggers T-cell activation—a double benefit[6].

Ebola

Our third infectious disease case study, the 2014/15 Ebola crisis, seemed to come from nowhere. After an initial slow response, it eventually energized the whole global life sciences community. This was not only to respond to the human need in West Africa, but also in part out of fear of the virus spreading beyond the countries of Guinea, Liberia, and Sierra Leone. From a single infected 1-year old child in Guinea who died in December 2013, the number of cases had grown to more than 25 000 by April 2015. Despite efforts to treat sufferers with the limited measures then available—maintaining fluid balance, oxygen tension, and blood pressure—the death rate reached 50%.

Although in many ways the 2014/15 outbreak showed authorities like the WHO and CDC as unprepared and sluggish, it also showed the medicines innovation process at its most accelerated. Some examples of exemplary responsiveness, on both sides of the Atlantic, included:

♦ TKM-Ebola, an experimental drug treatment developed by Tekmira, was granted US FDA 'fast-track' status to expedite its approval. This agent had actually been on FDA 'clinical hold' (suspension of human testing) as a result of safety concerns before the outbreak, and this constraint was released as part of a $140 million contract from the US government. The FDA's Expanded Access protocol to allow compassionate use of drugs pre-market approval has also been applied to TKM-Ebola.

♦ The FDA announced that, where necessary, the 'Animal Rule' (enabling the use of animal models rather than human trials) will be used in approval of treatments against Ebola.

♦ The EMA encouraged developers of Ebola treatments or vaccines to apply for 'Orphan Drug' designation. This designation ensures that further resources are provided by the EMA, including free scientific advice, fee waivers, and 10 years of market exclusivity post-authorization.

♦ The EMA also encouraged developers to apply in parallel to both the FDA and the EMA. Both organizations are working closely on sharing information, helping to spread resources and avoid duplication across countries.

At the time the outbreak occurred, one agent called ZMapp was already at a pre-clinical stage. This product was a mixture of three antibodies, manufactured in tobacco plants, from Mapp Biopharmaceuticals. It was rapidly advanced to clinical testing, with US regulatory authorities giving it clearance for human use under Investigational Drug Use (IND) regulations as early as February 2015. The US Health & Human Services department also exempted the company from legal liability.

However, rapidly scaling-up ZMapp production proved impossible, with the very limited supply being eked out on a few patients, largely for returning infected Western health workers. However, further anti-viral agents and candidate vaccines had been urgently brought forward, with two anti-Ebola agents (brincidofovir, from the US firm Chimerix, and favipiravir, from Japan's Fujifilm) going into trial in patients by November 2014. Blood plasma taken from recovered patients has also shown some efficacy and later proved a useful source of vaccine components.

Despite extreme difficulties handling the virus and mounting the clinical trials, such was the determination of all involved—industry, academia, global and national health agencies—that we may have a number of effective agents within a couple of years of the initial outbreak.

Clearly a vaccine would be even more important to control future outbreaks. A Canadian government-produced vaccine VSV-EBOV, developed by Merck, began a Phase 3 trial in Guinea in March 2015, vaccinating the friends, family, and neighbours of those already infected. The vaccine combined a fragment of the Ebola virus with a safe virus. First results are very promising, with 100% protection.

The lessons from Ebola

Need for global preparedness plans

This lesson needs to be learnt by the global health authorities, whose slow response showed them to be organizationally unprepared for a rapidly emerging threat like Ebola. It was, in fact, the Gates Foundation that moved most decisively, releasing $50m in funding by September 2014. *Charities can often be more nimble in funding urgent research.*

Adaptive development and regulatory flexibility

As we saw in H1N1, regulatory authorities, in response to real crises, are prepared to approve very shortened development programmes and give conditional approval to agents with very limited evidence. *Conditional approval is a viable route to save lives.*

Cross-stakeholder collaboration

Collaboration, especially between the public and private sectors, has also been a feature of the crisis. A Wellcome Trust/CIDRAP report ('Fast Track Development of Ebola Vaccines') urges 14 principles to guide such collaboration in the future[7]. The identification in advance of the 'target product profile' is one of these. *Collaboration, again, is energized by crisis.*

Pricing and reimbursement

It is also interesting that in crises such as this, there is little or no debate about product pricing. Indeed the release of substantial government funding and the willingness to spend $1m for the treatment of two patients in a US health centre shows that 'situations alter cases' when it comes to health products. Ebola is one of those cases in which we need a creative reward mechanism, such as an Advance Purchase Commitment (APC) to ensure sufficient research and product development between outbreaks that the next can be rapidly combated with a tested product. *Novel reimbursement models are needed for non-economic markets.*

We now come to a case study that shows the global life sciences community apparently powerless in the face of a threat that may be greater than any of the three reviewed so far.

Anti-microbial resistance

Antibiotics are, of course, one of the historic triumphs of medical innovation. From Fleming, Florey, and Chain's work on penicillin onwards, there were waves of successful innovation that developed increasingly powerful, broad spectrum antibiotics to bring under control the majority of infections in the developed world. However, the microbes have fought back: their rapid multiplication, and hence mutation rate, and their ability to exchange genes, coupled with the overuse and misuse of antibiotics, has created a set of resistant strains that now threaten to defeat the existing armoury of products.

In the face of this threat, and in sharp contrast to how mankind deals with the emergence of a novel infection, we have procrastinated about a danger that many think far more serious even than HIV, pandemic flu, or Ebola. If we don't tackle this problem, there are dramatic projections of future deaths from resistant infections—from the estimated three-quarters of a million today to 10 million a year by 2050.

The problem arises from two main sources. Firstly, *overuse* of antibiotics in human health, often for viral infections for which they are irrelevant. By the end of their first year of life, American children on average have had two courses of antibiotics! Secondly, from *abuse* of these same agents in animal health (to prevent animal disease and promote growth). The net effect of this dysfunctional use of antibiotics is the development of resistance in many classes of organisms. Resistance emerges remarkably rapidly: in the first use of streptomycin for tuberculosis, resistant strains appeared within the first courses of treatment.

The fact that bacteria are primed to some extent to resist antibiotics was discovered when the microbiome of an isolated Amazonian tribe, the Yanomami people, was studied. Although they had never been exposed to antibiotics or food from antibiotic-treated animals, their bacteria already had half a dozen resistance genes, though they were not switched on[8].

Modern antibiotic use both switches on resistance genes and also encourages the emergence of genomic variations that confer resistance. These genes are then exchanged between organisms to spread resistance across the global microbiome, for example, the ability to break down beta-lactams, via variants of the enzyme beta-lactamase[9]. A recent database lists 20 000 potential resistance genes in nearly 400 types of bacteria[10].

The result of all this exposure to antibiotics is the creation of infectious organisms that do not respond to common antibiotics and in some cases to none at all. Many of these organisms emerge in hospitals, and these nosocomial infections are amongst the most troubling (see Box 4.1).

The lack of convenient and speedy diagnostics compounds the problem. Often the identification of the specific organism awaits a process in which bacteria are streaked on a Petri dish on to which different antibiotic disks are placed. The ability of each antibiotic to stop the growth of the bacteria can then be assessed. However, this process typically takes days, not the minutes needed if it is to change the initial prescription.

Genomic testing, in principle, represents a basis for much faster determination. But so far too many of these PCR-based tests still take a number of hours and are intended to be run in central laboratories, rather than at the point of care, where a quick response could change the initial antibiotic choice.

Box 4.1 Nosocomial infections

There is a growing list of exotically named but deadly infections now circulating in hospitals: *Acinetobacter baumannii, Burkholderia cepacia, Campylobacter jejuni, Citrobacter freundii, Clostridium difficile, Enterobacter* spp., *Enterococcus faecium, Enterococcus faecalis, Escherichia coli, Haemophilus influenzae, Klebsiella pneumoniae, Proteus mirabilis, Pseudomonas aeruginosa, Salmonella* spp., *Serratia* spp., *Staphylococcus aureus, Staphylococcus epidermidis, Stenotrophomonas maltophilia* and *Streptococcus pneumoniae.*

Multiple drug resistant staphylococcus aureus (MRSA) and *Clostridium difficile* have become a source of many deaths. MRSA is an organism very closely associated with humanity (it appears in boils) and resistance to penicillin appeared early, via the development of the penicillinase enzyme, so that methicillin was developed in 1959 to evade this resistance mechanism, but methicillin resistance has appeared, and the organism has since moved outside the walls of the hospital to become a community-acquired infection.

C. difficile is found as an intestinal infection and it is believed that the widespread use of broad spectrum cephalosporins has depleted the microbiome levels of Gram-negative bacteria. This has given more opportunity for this Gram-positive infection—which is spread easily by hospital staff and equipment via its spores[a].

[a] Davies J and Davies D (2010). Origins and Evolution of Antibiotic Resistance. *Microbiology and Molecular Biology Reviews*, Volume 74, Issue 3, pp. 417–433.

However, even if the determination of the microbial variant can be made fast enough, we are running out of therapeutic options. The key barrier to developing fresh classes of antimicrobials to combat resistant organisms is the lack of adequate incentives. This problem arises in part from the low cost of existing generic antibiotics with which they might be compared, but the tougher problem is that the right use of the new generations of agents will be restricted to relatively rare cases of extreme resistance. Therefore the markets, measured conventionally as volume times price, are not economically attractive to pharmaceutical companies. The lack of such incentives means that we have developed no new classes of agents over the last 20 years.

This is not just a problem for those who enter the physician's office with an infection. Antibiotics are an integral part of hospital surgery (as well as combating nosocomial infections—see Box 4.1). Much routine hospital care would become high risk if no effective antibiotics were to hand.

We need new types of incentives, such as advance purchase commitments (pioneered for another case of non-economic markets: vaccines for neglected diseases). Another alternative is fixed value prizes. Yet another is rewarding successful development of new agents with some kind of transferable 'value' that is usable in other settings, such as longer patent protection or fast review vouchers for normally economic products.

This could extend to creating vouchers that a small company without non-antibiotic products could sell to a larger company to realize their value. In all such mechanisms we 'de-link' the reward for the developer from the volume that might result from conventional promotion and widespread use—which would of course defeat the objective of creating 'reserve' agents for resistant cases only.

Following concerns highlighted by its Chief Medical Officer, Sally Davies[11], the UK government commissioned a report by a leading financial sector economist, Jim O'Neill, on the AMR phenomenon and recommendations to address it. One of the latter was that the global pharmaceutical industry should contribute to a $2bn fund to research new antibiotics: it remains to be seen whether industry sees AMR as their problem to fund, rather than the health systems themselves[12].

What do we learn from this so-far rather unsuccessful innovation experience?

Lessons from antimicrobial resistance

Product over-use

In infectious diseases, product use needs to be selective in a way that does not apply to non-infectious conditions. Organisms adapt and remember, exchange and propagate their adaptations. Hence more precise targeting is vital. *Greater precision implies an element of control over prescriptions.*

Role for rapid diagnostics

We need diagnostics that can give the doctor the basis for this better targeting in minutes, not hours or days. Otherwise the need (or temptation) to start the patient on a broad-spectrum anti-infective will further worsen the problem. *Effective, convenient diagnostics are central to precision medicine.*

Limitations to market mechanisms

The normal operation of the market for innovative products does not always serve our needs. In this case, should a novel antibiotic be developed and widely promoted it would itself become subject to the development of resistance, defeating the goal. (This is not the only area in which this lesson is important: any stratified or precision medicine faces the possibility of use in patients for which they are not suitable). Other mechanisms than conventionally rewarded commercial R&D need to be set up when the normal operation of markets does not deliver what we need. The use of public–private consortia may be especially appropriate when markets are atypical or marginal in economic value.

When needed innovation is not forthcoming, we need to question the incentives and redesign them if necessary.

The next few years will demonstrate whether the necessary creativity—and especially the political will to set up novel payment mechanisms—can confront this global challenge. The risk is that individual countries will prevaricate and fail to collaborate together in sufficient numbers to create a pipeline of products and a reward mechanism that works.

Cystic fibrosis

Cystic fibrosis is a deeply distressing condition for both the children it afflicts and their parents. Increasingly, however, we understand the underlying genetic mutations that cause it and how these vary from patient to patient. The mutations are in the genes coding for a trans-membrane conductance regulator that controls chloride ion transport across membranes, including those in the lungs and sweat glands (the latter giving rise to an easy 'chloride in sweat' diagnosis and response test for any potential drug).

However, until quite recently the only drugs available for the condition treated the symptoms and sequelae of cystic fibrosis, including lung infections and GI insufficiency. There were no treatments to deal with the condition itself.

Cystic fibrosis is an autosomal recessive disorder—in other words sufferers have to inherit two copies of the defective gene, one from each of their parents. We have now identified more than 1500 mutations in the gene for CFTR, on chromosome 7. The most common, affecting two-thirds of patients worldwide, is a three-nucleotide deletion resulting in a loss of one amino acid at position 508. However 4–5% of patients have a substitution at position 551, and it is this group of patients that initially benefited from a breakthrough drug called Kalydeco (ivacaftor), from Vertex Pharmaceuticals.

This mutated form of the protein migrates to the cell surface as it should but fails to transport the chloride ion. Ivacaftor binds to the protein and opens the channel. It has recently been approved for the *R117H* mutation also, only present in 500 sufferers in the USA, on the basis of a trial involving 69 of those patients. The subsequent extension of its use for mutations *G178R, S549N, S549R, G551S, G1244E, S1251N, S1255P*, and *G1349D* was based on a 39-patient study on patients selected for these even rarer mutations[13]. A longer-term study of safety in those previously involved in a trial is now underway[14].

This innovation resulted not only from the efforts of Vertex but also from the active involvement from the US Cystic Fibrosis Foundation, headed by the dynamic Bob Beale. The organization played a major role in development, in funding research to the tune of $150m, and in advocacy for its approval and use. It was also a matter of justified pride at the FDA that the agency examined and approved the drug on the basis of trials in so few patients, and in just 3.5 months[15].

However, the payers were less enthusiastic, since the price charged by Vertex is $307 000 per year per patient, Despite the drug targeting only 4% of the cystic fibrosis population, the worldwide annual sales are already more than $500m.

Interestingly, the Foundation that initially co-funded the work has sold the royalties in 2014 for $3.3bn and is applying the proceeds to funding further research with Vertex and other pharma companies. It is therefore acting as a venture capitalist, not just a disease charity, and this 'venture philanthropy' lesson is being taken up by other charities, such as the Multiple Sclerosis Society[16].

The company has now gone on to develop a product called lumacaftor which is being trialed with ivacaftor in patients with the commonest *F508del* mutation.

For a summary of the timeline and key facts, see Figure 4.1.

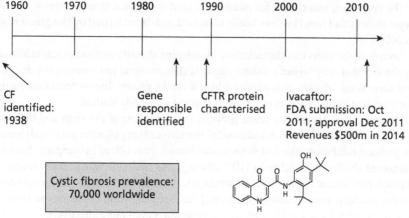

Fig. 4.1 Cystic fibrosis—timeline and key facts.

Lessons from ivacaftor for cystic fibrosis

Disease mechanism research

Without deep molecular understanding of the cystic fibrosis disease mechanism, in terms of the genes implicated and their effects on protein structure and function, the drug could never have been developed. *Disease mechanism is the starting point for most R&D.*

Patient organization involvement

This case study reinforces the value of an energetic patient organization advocating research and the approval and use of an effective product. It goes further, to show *patient organizations can be effective research funders.*

Regulatory flexibility and speed

The willingness of regulators to rapidly approve on the basis of trials with limited patient numbers—and to use the cumulative safety information acquired in these trials—to make further trials in rarer populations a feasible proposition. *Regulators are vital partners in clinical development design.*

Hepatitis C treatment

HCV is a huge health burden worldwide: it is estimated that 3% of the world's population is infected. It is often detected late, since early symptoms are few and therefore others can subsequently be infected via sexual transmission, or contaminated needles (and, at one point, unscreened blood transfusions). Over the long term, infection leads to liver cirrhosis and ultimate failure resulting in the need for costly transplants and often the development of liver cancer. Many years of high healthcare costs result.

Some treatments to keep the virus at bay emerged around 20 years ago, and until recently the best regimen was interferon-A and ribovarin, with its flu-like side-effects and inability to eliminate virus from the patient's system.

The virus itself was elusive for many years (and the disease termed 'non-A, non-B' hepatitis over that time) but was finally identified and characterized by Houghton et al. in 2009[17].

Even once the virus was characterized, developing effective anti-virals was inhibited by the fact that only higher primates (such as chimpanzees) can contract the disease, and there is no cell culture system in which it can be grown. The proteins making up the virus had to be cloned in other organisms to be properly studied.

However, over the last few years, growing understanding of the virus and the proteins that propel its replication has now led to a series of very effective anti-virals based on protease inhibition. The first of these was Sovaldi from Gilead (a company that had pioneered similar approaches in HIV, although this molecule was actually acquired through purchase of the company Pharmasett). The drug (and a small number of competitor products now reaching the market) has been shown to eradicate the virus, a feat as yet not achieved in HIV, where the virus hides even more effectively.

While clinically the drug has been warmly welcomed, and health technology agencies have deemed it cost-effective, there has been wide controversy about the price (over $80 000 per course of treatment in the USA). Because of the prevalence of the infection, the bill for the drug has been seen as unaffordable in many countries, or has been restricted to very severely affected patients. While the latter is understandable, it goes against the commonsense drive to treat patients early, before severe symptoms appear and the virus is transmitted to others.

Now that potential competitor products have appeared, payers have predictably begun to take very aggressive tendering or negotiating stances, and it seems unlikely that the initial price level will be sustained. In some ways Sovaldi marks a watershed in the approach by payers, particularly in the USA, which has so far largely been prepared to afford high drug prices.

Gilead has offered the drug at dramatically reduced prices in countries with high HCV burdens but much lower GDP/capita levels.

Lessons from HCV/Sovaldi

Understanding viral biology and life cycle

The detailed knowledge built up in the last 10 years has eventually yielded effective products, as we saw in HIV. *Again, there is no substitute for basic biological understanding of disease.*

Innovation often comes through the SME route

Although Gilead had a very active antiviral R&D programme, the product came from a smaller company. This pattern is repeated frequently and has implications for how the life science industry develops.

Reimbursement and affordability

This example shows that affordability is just as important to most payers as long-term cost-effectiveness, especially when the disease has a high prevalence. *Novel reward mechanisms are needed for high initial cost/long payback therapies.*

Neonatal diabetes

Diabetes treatments and their evolution hold many lessons, but some specific forms of neonatal diabetes (diabetes diagnosed at less than 9 months of age) are already good examples of precision medicine in action. The more common Type 1 and Type 2 diabetes are genetically diverse, with many genes contributing to risk. However, some forms of neonatal diabetes (less than 5% of juvenile diabetes cases) result from very specific monogenic mutations. One of these is in *KCNJ11*, a gene coding for a protein that controls a potassium ion channel in pancreatic beta cells. In normal individuals, this channel opens in the presence of blood glucose to release insulin and so bring down glucose levels. In patients with one of 30 mutations in the KCNJ11 gene, the channels remain closed. Their pancreatic islets manufacture plenty of insulin but this is not transported out of the cell.

Researchers demonstrated that an already available oral drug, a sulphonyl urea, could bind to the channel and enable release of the insulin. Patients could then avoid the distressingly frequent insulin injections on which Type 1 diabetes sufferers normally need to rely.

While this is another story of successful research into the genetics of disease and also one of the 'repurposing' of existing drugs, one specific patient case shows how knowledge can spread among engaged patients (or in this case parents) in the digital age.

When only 5 months old, Cameron Lundfelt had been diagnosed with juvenile diabetes and so the family confronted the challenge of lifelong insulin therapy. They lived in Alaska, where there was at that time no paediatric diabetologist. As insulin wasn't controlling his glucose as it should, and Cameron was suffering developmental problems, the family consulted a string of other specialists. Cameron's mother—anxious to do all she could for her son—went to the Web and saw the story of Lilly Jaffe, a child being treated by Lou Philipson, a specialist in Chicago. He conducted a gene test on Cameron—via the mail—and it confirmed a mutation in the *KCNJ11* gene of a very rare kind: a colleague Andrew Hattersley in the UK had identified just three cases similar to Cameron's.

Lessons from neonatal diabetes

International research collaboration

The case shows the power of piecing together the genetics of very rare diseases by the efforts of specialists in international networks. *Rare disease research must be transnational.*

Distributed knowledge

In the age of the Web, transparent patient-friendly information can be a great adjunct to such a medical network. There is no-one with a greater incentive than the patient and family. While not every part of the globe can have specialists in rare conditions, their *knowledge can be globally available.*

Precision medicine will rely on both: very specific molecular diagnosis and the ability to apply it in individual circumstances despite the challenges of geography.

Motor neuron disease

Three names, but one disease. In Europe, the progressive death of motor neurons leading to decline and typically rapid death, is often called motor neuron disease (MND). In the USA, it is called amyotorphic lateral sclerosis (ALS) or Lou Gehrig's disease (after the Major League baseball player who contracted it). Whichever of these names the neurologist uses, the diagnosis is equally dire.

The disease was first described more than 140 years ago but little was understood of its cause until recently. Genetics have now linked ALS with mutations in two genes: one coding for the enzyme superoxide dismutase, the other that codes for valosin-containing protein (VCP), involved in the breakup and clearance of proteins via RNA granules. As often is the case, these insights came from studying families with a higher propensity to develop disease[18].

The protein breakdown linkage is promising, but the pathways involved are very complex and the VCP mutations can emerge as ALS, as fronto-temporal dementia, or as Paget's disease of the bone! Designing intervention points in the complex system of protein homeostasis is far from straightforward.

ALS was the diagnosis faced by Les Halpin, who was a statistician and leading UK businessman. Knowing that the last new medicine had been approved 20 years previously and that it offered only a marginal benefit, Les started a movement called Empower, for earlier and easier access to experimental treatments. This movement was instrumental in encouraging the UK government to introduce its 'Early Access to Medicines' scheme in 2013: a telling legacy for an inspirational individual.

However, the disease that motivated this initiative is not yet itself fully ripe for such an approach. The insights we are developing into the mechanism for the decline and death of the patient's motor neurons are still too recent. So while MND provides a story of strong patient activism, it also highlights where the life science community should focus—on further study of the molecular pathology of the disease and exploring therapeutic routes.

Lessons from ALS

Disease mechanism research

In a complex degenerative disease, there is no substitute for detailed work on mechanism and without this, designing treatments is essentially impossible. Further support is needed for *more research on disease mechanisms.*

Patient empowerment

Patients suffering from such diseases, however, have the right to be heard and to try any promising therapies. Even patients that cannot themselves benefit can help change the innovation climate, as did Les Halpin. *Patients can move policies.*

Rheumatoid arthritis

Rheumatoid arthritis (RA) is a potentially crippling disease of auto-immune origin. It is as old as mankind: evidence of it is seen in Egyptian mummies and it was first described in Indian literature around 250BC. Effective therapies of any kind had to wait until the mid-20th century, with gold and cortisone having some impact, and in the 1980s the development of methatrexate (which is still the initial drug for RA).

Also in the 1980s, research on the cytokines found in rheumatoid joints discovered the role of tumour necrosis factor (TNF), beginning the search for molecules with anti-TNF activity. They were first trialled by Marc Feldmann in 1992, and were found not only to improve symptoms but also halt joint damage. That is, for the first time they were 'disease-modifying'.

The result is one of the most striking triumphs of drug therapy, literally emptying wards of patients waiting for joint surgery. The mechanism of the new class of drugs seemed clear, and the timing was right, since we had learned how to make the re-engineered antibodies that were effective as TNF-alpha inhibitors on a commercial scale. The first successful biological product was Enbrel, from Immunex, a fledgling biotech company in Seattle, which became part of Schering-Plough and subsequently Pfizer. A variety of products emerged over the next few years and the most popular of them, Humira from Abbvie, became the world's highest revenue drug (Figure 4.2).

Despite the multiplicity of products, the RA challenge is far from fully conquered. Around a quarter of patients do not respond to these biologicals, for reasons we do not yet fully understand. For other patients, a series of products need to be tried before the most suitable is found. Some of the drugs cause immunogenic reactions in some patients.

The success with the products also spilled over into other auto-immune diseases such as psoriatic arthritis and psoriasis, and also into the treatments of many cancers.

In the 2010s, oral small molecule drugs began to appear that rivalled the impact of the TNF-alpha antibodies but avoided the painful injections and slightly increased risks of cancer and infection associated with the suppression of the immune system.

While the biological and clinical triumph became clear, it was some time before the HTA system recognized that the revolutionary effects of the drug were also saving the health system substantial amounts of money. They were much more expensive than the very cheap generic methatrexate that had been first-line therapy for many years.

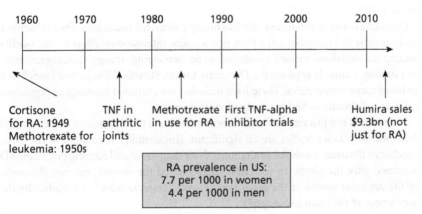

Fig. 4.2 Rheumatoid arthritis—timeline and key facts.

Hence, when NICE first appraised the drugs it failed to approve them, to the frustration of rheumatologists. And even once the HTA agencies had given them the green light, payers remained reluctant and usage is still far from universal.

Lessons from rheumatoid arthritis

The power of the mechanistic approach to drug discovery

This, combined with the recombinant DNA ability to produce large amounts of antibodies, fulfilled their promise, transformed millions of lives and created a market of many billions of dollars.

The challenge of initial health economics

The difficulty in understanding the economic impact on the whole pathway of care caused the system to initially undervalue the advance. RA is not immediately life threatening. Its consequences: slow deterioration, a loss of ability to work, and increasing incapacitation, are not as compelling to health systems as some more immediately tragic diseases, for which higher cost per QALY thresholds are sometimes applied. RA is an example of a disease whose treatment revolutionizes the whole pathway of care and so requires a 'value pathway' approach to HTA. *Health economics needs to develop this 'value pathway' methodology.*

In this discussion of arthritis, it is important to ask why progress in RA has not been matched by progress in its auto-immune cousin, osteoarthritis, which can also cripple, particularly in the elderly.

Dementia

This devastating disease is not only a major health threat to the increasingly large numbers of people entering their later years, it threatens the very sustainability of our health and social care systems. Around 47.5 million people are estimated by the WHO to suffer from the disease, although only around 15 million have a formal diagnosis. This number expected to reach 75 million by 2030, as populations age. Current worldwide dementia costs are estimated at more than $600bn, or 1% of global GDP.

Dementia is a global challenge that has finally galvanized politicians. The G8 Summit on Dementia in December 2013 launched a major international effort by the world's leading industrialized nations to collaborate on developing strategy, funding research, and finding a cure. It appointed a Dementia Envoy, Dennis Gillings, and launched a series of major conferences. These have included international meetings of regulators ready to collaborate to find better development routes.

Dementia is not just one disease. Alzheimer's disease (AD), vascular dementia, and dementia with Lewy bodies are all significant. (Incidentally, the names of the three conditions illustrate nicely the limitations of our diagnostic and naming practices: AD is named after the physician who first characterized the disease, vascular dementia by the apparent source of the brain lesions, and dementia with Lewy bodies by the appearance of the brain in autopsy!)

We are learning that some lifestyle changes can avoid or slow dementia. Vascular health, important to so much of our well-being, seems to be a major factor in dementia also, as the recently published FINGER study demonstrated[19]. However, prevention and stimulation can only be parts of the answer. We need treatments, and no really effective treatments have yet emerged from a huge R&D effort.

The only drug treatments that are currently available are the so-called anti-cholinesterase inhibitors, drugs that keep the level of choline—a vital neurotransmitter that carries messages from one neuron to the other in the brain—as high as possible. But these do not reverse the underlying mechanisms of decline, merely help the patient's brain make best use of what remains. And there have been no new drugs launched in over a decade.

There seems to be a cascade of molecular events leading to the demented brain, taking Alzheimer's as the example. A protein circulating in the blood, beta-amyloid, begins to aggregate and these aggregates build up in the brain. Next they start to form tangles and plaques, which are accompanied by the buildup of a second protein, called tau, in a heavily phosphorylated form. Then the neurons start to die—or rather the normal death rate of neurons rapidly increases, leaving whole areas of the brain without grey matter.

Despite intensive study of the disease process—through careful examination of post-mortem brains and in vivo study via CT, MRI, and PET imaging—we still do not fully understand this sequence of events. We still cannot separate causes from effects. There are even some theories that AD is virally transmitted, as a result of the typical pattern of invasion of the disease into the brain.

The most obvious feature of the disease—the tangles and plaques of aggregated amyloid—have also been the most obvious target of drug research, aimed at reducing the level of amyloid and/or slow or reverse the aggregation process. This belief that targeting amyloid will lead to a dementia treatment—the 'Amyloid Hypothesis'—has been the dominant approach until quite recently.

The results so far have been tremendously disappointing. In the last decade, no fewer than 1120 drugs have entered the pipeline. None have emerged successfully from it. Pfizer, J&J, Roche, and Lilly have invested heavily in potential products based on the amyloid hypothesis but all have failed to make it through development. These high attrition rates, coupled with difficulty recruiting patients for trials, makes this the most costly area for pharmaceutical research[20].

Most of the development failures can be attributed to seeking to treat patients with fully developed AD, for whom much of the brain decay and function loss has already occurred and is unlikely to be reversed. It has become clear that we need to treat patients at much earlier stages in the disease.

We have recently seen a significant set of more positive results from study of an antibody, aducanumab, from Biogen (see Box 4.2).

Success of anti-amyloid agents, both in trials and in potential future routine treatment, will depend on careful patient selection. Work underway by Simon Lovestone in Oxford on a blood marker of early disease (followed by tests of cerebrospinal fluid and brain scans) is one approach.

Box 4.2 Finally a new treatment for Alzheimer's?

Aducanumab targets aggregated forms of a-beta, the circulating precursor of amyloid. Interestingly, the antibody was originally derived from aged patients without dementia, with the hypothesis that they may have an antibody helping keep dementia at bay. The 56 patients in the Phase 1 trial were mild to moderate sufferers and were monitored by brain scans over 2 years and showed no increase in amyloid-related imaging abnormalities (ARIAs). Very high doses (equivalent to 4 g for a 150 kg individual) were shown to be safe and so a Phase 2 trial was started on 166 subjects with prodromal or mild AD, with the beginnings of amyloid buildup, with intensive PET and MRI imaging throughout.

Results recently presented by Biogen showed dose-dependent amyloid reduction in the brain. More importantly, the drug seemed to reduce the rate of cognitive decline as measured by the MMSE (mental acuity) and CDR-SB (ability to function) methods. These results were sufficiently positive that a Phase 3 study began in December 2014, expected to report in 2019. So what seemed as a research avenue that was becoming a dead end is still being pursued, both by Biogen and also by competitors Roche and Lilly, who have antibodies with similar characteristics.

Of course, it may still be true that other mechanisms of action prove more successful: phosphorylated tau is a target of a Phase 3 programme from the Singapore- and Scotland-based company TauRx. Here the subjects are fronto-temporal dementia patients. The drug, derived from the dye methylene blue, is called Rember!

Regulators are poised to be very cooperative. The FDA has approved two PET imaging agents to assist the measurement of plaques and has also published guidance on designing trials and assessing therapeutic impact in early stage patients[21].

Recognizing that innovation may come from academia or SMEs, two companies (Eisai and Lilly) plus the UK Medical Research Council and Alzheimer's Research UK have created a public–private consortium to support dementia drug developers[22]. Companies are also considering running Phase 3 studies in parallel.

Any effective treatment for AD, unlike most future medicines in the precision medicine era, is likely to be a true blockbuster. However, it would not be without ethical and practical issues if—as seems likely—it needs to be taken before the development of obvious symptoms.

Lessons from dementia

Disease mechanism

We have already seen the challenge of developing treatments in the absence of a very clear understanding of disease mechanism. While we know that amyloid build-up is an important feature of the disease, we still lack definitive evidence that stopping it will halt the disease. This will only come from trials such as Biogen's.

Even so, we must remember that *dementia, like many conditions may have diverse causes.*

Predictive markers

Predictive diagnostic biomarkers to select patients for trials are clearly essential to progress in AD. Such markers will also be vital to persuade pre-symptomatic patients to take a prophylactic drug, once it is developed. In this disease, like many others, *effective diagnostics are as important as—and usually precede—therapeutics.*

Collaboration

Pre-competitive cooperation between companies, and perhaps parallel clinical trials, would share the substantial investment of drug development and make a challenging field more attractive. *Collaboration can overcome economic barriers.*

International regulatory action

The Dementia Challenge has surfaced the *willingness of regulators to collaborate* together in the face of a major global priority.

Ulcer treatment

The effectiveness with which we can now treat ulcers, without highly invasive stomach surgery and with the prospect of complete cure, is one of the major successes of drug therapy in recent decades. From a therapeutics point of view, this is a three-part story.

The first part was the emergence of histamine H2-antagonists first explored in the mid-1960s and launched a decade later. Knowing that histamine was involved in acid production but that traditional anti-histamines were ineffective, James Black and his team at Smith, Kline and French sought a second set of histamine receptors and made literally hundreds of histamine-like molecules in an attempt to block its action in the stomach. The best drug that emerged, Tagamet (cimetidine) became the first 'blockbuster'. The successor drug, Zantac (ranitidine) from Glaxo was even more successful.

However, the pharmaceutical industry did not rest there, and sought other, even more direct, mechanisms to inhibit stomach acid production. Prilosec (omeprazole, from Astra) was the result. It is a selective and irreversible 'proton pump inhibitor' inhibiting the enzyme system on the surface of the stomach's parietal cells. This product, too, had a successor in the form of Nexium (from the same company): equally successful, if not much differentiated.

But, asked two Australian researchers, Barry Marshall and Robin Warren in the early 1980s, why did the ulcers begin in the first place? They discovered the bacterium *Helicobacter pylori* in patients with gastric ulcers, This flew in the face of the prior belief that bacteria could not live in the stomach's acid environment. Warren even demonstrated the bacterium's causative power by swallowing it himself and triggering acute gastritis! His and Marshall's Nobel prize in 2005 was doubly earned.

So now ulcers are usually treated with combinations of two antibiotics (e.g. amoxicillin and clarithromycin) and a proton pump inhibitor ('triple therapy'). They remain a serious condition, if not quickly diagnosed but—once so—can be readily treated with pharmaceuticals rather than stomach surgery. For the overall timeline see Figure 4.3.

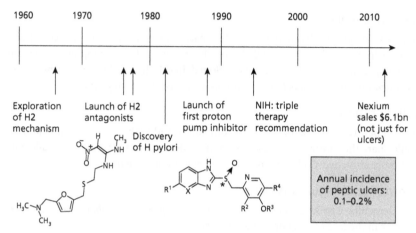

Fig. 4.3 Ulcer treatment—timeline and key facts.

Lessons from ulcers

Mechanistic persistence

Even if there appears to be a reasonably effective therapy, digging deeper into the causative biology can open up new and more effective treatments. *A succession of new mechanisms of action makes for broad therapeutic strategies.*

Combination therapy

This is often more effective than monotherapy if the effects are synergistic.

Bold investigators

In former times, drug developers would often conduct the first trials on themselves. This is the only example I know of a bold investigator giving himself the disease first.

Multiple myeloma

There are two main classes of therapy for this blood cancer. They each hold interesting but quite different lessons.

The first are the 'imids'. The first 'imid' was a drug with an extremely chequered past—thalidomide. Thalidomide is a drug whose name is known more widely, at least among a certain age group, than almost any other. Patented by Grunenthal in 1954, it was licensed as a sedative and sleeping pill in 1957/58 in the UK and much of the rest of the world. Sadly, it was also prescribed to pregnant women with morning sickness.

Thalidomide was used quite widely in Europe for the next 3–4 years, and was found to cause terrible birth defects in more than 10 000 children, if pregnant women took it. Of the 2000 affected babies in the UK, only 466 survived. It was withdrawn in 1961, but—as Peter Lachmann, the author of 'The penumbra of thalidomide, the litigation culture and the licensing of pharmaceuticals' puts it—it has cast a very long shadow ever since[23].

It was not licensed in the USA, where the newly appointed FDA reviewer, Frances Kelsey (for whom this was her first dossier) asked for more pre-clinical toxicology studies. It is felt by some that the greater conservatism implied by that obviously correct decision has encouraged the FDA ever since to err on the side of safety. This is despite the fact that such studies would not have been conducted on pregnant animals in any event. (Scientists then believed that drugs would not pass the placental barrier: we now know that many things can.) The FDA even gives staff members 'Kelsey' awards, which ensures they do not forget their duty to protect the public: it recently gave the first such award to Frances Kelsey herself at the age of 96!

This one episode gave rise to much of the regulatory machinery we have today, including tests to specifically exclude the 'teratogenic' effects found in thalidomide. The US Congress passed the Kafauver-Harris Drug Amendment to require manufacturers to provide 'substantial evidence' of effectiveness and safety and to give the FDA oversight of clinical trials.

Understandably, it was many years before any researchers or doctors dared to investigate the molecule as a potential therapy. (Laboratory studies continued to discover the mechanism of the drug's devastating effect.) However, in 1964, an Israeli physician Jacob Sheskin found it to be a potent drug for erythema nodosum leprosum, a form of leprosy, for which it was approved in 1998. It was subsequently found to be a potent suppressor of the runaway growth of mutated B cells in multiple myeloma patients and licensed for that indication in 2006.

Two chemically related 'imid' drugs, that have largely replaced thalidomide itself, are now major clinical and commercial successes, with all three being marketed by Celgene. But their use is very closely controlled by a 'risk management system' to avoid any risk that a pregnant woman ever takes them again.

We now understand both the mechanism for the devastating negative effects of this class of drugs and for their life-extending power. They bind to a protein, cereblon, one that plays a crucial role in protein homeostasis—in particular the process by which proteins are tagged for removal. The process turns out to be important for the proliferation of myeloma cells in multiple myeloma, as well as in embryonic limb formation.

A second drug that has made a major difference to the lives of many multiple myeloma patients is Velcade (bortezomib). It is a proteasome inhibitor, also involved in protein degradation. It was originally synthesized in 1995, taken into clinical trials in October 1999 and approved by the FDA in May 2003—fairly rapid progress.

Unlike some cancer therapies, multiple myeloma drugs deliver substantial life extensions, and their manufacturers therefore expect this value to be reflected in price. However, when Velcade first came to the market in the UK, in the early days of NICE, the calculated cost per QALY was over the £30 000 threshold that NICE normally set, and reimbursement was refused.

Johnson & Johnson, who were marketing the drug in the UK, came up with an ingenious response. The normal test for drug response in patients was measurement of a protein called 'M protein' in serum. They proposed that drug given to patients that did not achieve at least the minimum response rate would not be reimbursed. The company estimated that this would reduce the cost to the NHS by 15%. On that basis the drug's reimbursement was approved[24].

This was one of the earliest of the so-called 'risk-sharing schemes' pioneered in the UK but also taken up in the USA. Although these latter arrangements tend to be confidential, it is known that one company marketing and an osteoporosis drug (for bone loss) has offered to pay for fractures in patients consistently taking the drug; another with a diabetes drug has linked its reimbursement to adherence and glucose control[25].

Myeloma has also seen some of the most proactive strategies on the part of patient organizations. Myeloma UK actually runs clinical trials, as well as very actively engaging with regulators and HTA agencies. They seek to define the future treatment pathways for the disease, including specifying diagnostics to more effectively stratify patients for trials and treatment.

Lessons from multiple myeloma

Mechanistic insights

Drugs that modulate fundamental mechanisms can have both positive and negative effects that seem quite unrelated: we are still piecing together much of the relevant biology in the case of thalidomide. *Drugs can be potent tools in probing biology.*

Drivers of regulation

A major drug disaster has a severe inhibitory effect on the development and regulatory system as a result of a 'never again' reaction. This can be both in the form of specific new tests introduced and also an overall culture of caution. *It is vital to learn the real lessons of any major drug withdrawal.*

Risk group exclusion

Known side-effects can, however, be very effectively excluded if the 'at risk' group can be clearly defined (in this case, pregnant women). We now have very effective models of REMS (Risk Evaluation and Management Systems) to deploy when needed.

Payment creativity

In the case of both drugs in the UK, the innovator company proposed a mechanism to reduce the risk and outlay on the part of the health system. Such proposals may perhaps pave the way for arrangements in which the manufacturer only gets paid when the product works. However, this approach relies on our ability to easily track outcomes and administer rebates: the *practical difficulties of tracking product use and linked outcomes have been the major barrier to a wider adoption of risk-sharing schemes.*

Patient engagement

Myeloma has seen very active examples of this. Some patient organizations are ready to be taken into full partnership in the innovation process, co-sponsoring clinical trials.

Gleevec: the first truly targeted cancer therapy

Chronic myeloid leukemia (CML) was the first cancer to have its genetic and metabolic basis really understood. In 1960, Nowell and Hungerford[26] saw that chromosome 22 was dramatically reduced in size; it was dubbed the Philadelphia chromosome, after the city where both researchers worked. By 1973, it had become established that an interchange of DNA between chromosomes 9 and 22 was responsible for the change[27]. Baltimore and Witte went on to show that a specific tyrosine kinase Bcr-Abl was erroneously coded after this interchange, and that this kinase was activated in CML. It was responsible for the uncontrolled proliferation of white cells in CML patients.

However, it was not until the mid-1980s that real research interest emerged in the possibility of selectively inhibiting Bcr-Abl. After all, there are 150 or so tyrosine kinases in the body, with the odds that a compound could be found that targeted just one in its ATP-binding pocket seemed remote. However a programme was started at the Swiss company Ciba-Geigy (subsequently merged into Novartis) that brought together talented oncologists, biologists, and a medical chemist called Jurg Zimmermann, who ultimately designed a molecule then called ST1571, in 1992.

It proved to inhibit phosphorylation, the step catalysed by Bcr-Abl and it was demonstrated a year later to kill CML cells. Then began the usual battery of pre-clinical tests, which brought disappointment: the potential drug had toxicity in dogs. Not only that, but it was very tough to make, requiring a 12-step manufacturing process. And the drug tasted bitter and had poor solubility (the latter is a death-knell for many potential products).

From the commercial standpoint, CML was a much less attractive oncology market, at 6650 new cases per year in the USA, than for example prostate (200 000) and breast (195 000) so marketing managers were unenthusiastic. However the company persisted. Following the merger that created Novartis, the new CEO Daniel Vasella admired the work that the team had done and chose to overlook the practical problems—fortunately, for both CML patients and Novartis shareholders.

The drug went into Phase 1 trials in 1998 and by the following year had been shown to reduce to normal the white cell count of all 31 patients in the trial. Not only that, in many cases there was cytogenetic remission: the Philadelphia chromosome partially or fully disappeared. Phase 2 was quickly initiated the following year and showed similar results. An expanded access programme increased the total population on the drug to 2000.

After that, the pace quickened further. In July 2000, the FDA granted fast-track designation, the New Drug Application was submitted in February 2001, and given priority review status the following month, so that by May of that year the Health & Human Services Secretary was standing on a podium, alongside Vasella, announcing FDA approval of the first truly targeted cancer drug. For once, the term 'magic bullet' seemed appropriate (see Figure 4.4).

This, of course, was just the beginning of the story. On the plus side, the drug was quickly shown to also have activity in gastro-intestinal stromal tumour and came under investigation for other tumour types, alone or in combination. On the downside, like most successful cancer drugs, resistance eventually emerged and Novartis and others have sought successor products to use after this resistance appears. The product has also been the focus of a very heated and unproductive battle between the company and the Indian government on the validity of the patent on the specific form of the drug used in the commercial product, and generic forms of Gleevec seem likely to appear in the US market by 2016.

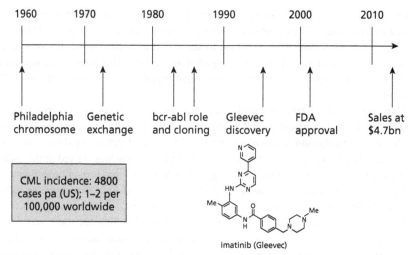

Fig. 4.4 Gleevec—timeline and key facts.

However, the current (2014) market for Gleevec (generic name imatinib) is $4.75bn, and it is Novartis' best-selling product. Because of its transformative power in a poorly served patient population, NICE approved it (for patients in the late, blast cell phase) at a cost per QALY of £49 000—the highest accepted up to that point.

Lessons from Gleevec

Market projections

Beware of condemning a potential drug on premature market analysis based purely on current market size or disease prevalence. Gleevec is not the only product that, once approved for an apparently niche indication has gone on to become a major drug. *Precision medicine does not necessarily lead to small ultimate markets.*

Mechanism of action

It was a long, but indispensable journey from the initial chromosomal observation to the drug target. *The knowledge about the underlying mechanism of CML was crucial in ultimate success.*

Clinical candidates

The next journey, from disease mechanism to clinical candidates, can be a major challenge and require *sustained clinical–biological–medicinal chemistry teamwork*, and managerial support.

Regulatory stance

When a drug is a genuine breakthrough, regulators can be both highly flexible and fast-paced: it would be hard to fault them in this case.

Patents

Whatever the nature of the science and the clinical impact, it will not stop some payers and generic competitors from seeking to overcome patent protection: the incentives are just too great.

Hypertension

There is not just one source of hypertension, but a large proportion of patients respond to one of a small number of classes of drug: beta-blockers, calcium channel blockers, angiotensin converting enzyme (ACE)-inhibitors, and angiotensin receptor blockers (ARBs) are the most common. This one clinical problem has probably generated more pharmacological activity than any other. The story, however, shows how steady scientific effort over decades can overcome both T0 and T1 problems.

Since the risk of heart attacks, strokes, and other serious conditions rises steadily with blood pressure, there is a need to ensure that patients adhere to therapy—despite not feeling any immediate benefit.

An area of Cambridgeshire in the UK was the subject of an experiment involving a new technology, called Proteus. The pills are coded with a radiofrequency tag that is 'read' by a patch on the patient's body. Via signals from the patch, researchers (and prescribing doctors) can know when and to what extent patients comply with the therapy they are prescribed.

The results were interesting: some patients that doctors might have assumed had varieties of hypertension that did not respond to drug therapy were not responding for a simpler reason: they weren't taking the tablets–or at least not consistently enough[28]. Research continues to understand why particular patients don't comply: is it real experience of side-effects, fear of them, or simply a failure to appreciate the risks that high blood pressure brings?

The *lesson* here is that, if we can bring digital technology into the process of drug adherence we will both learn more of the thought process of patients and also improve the performance of prescribed therapy.

Most of the case histories so far are of pharmaceuticals. We now look at the neighbouring sectors of vaccines, diagnostics, and devices to capture learnings from across the life sciences spectrum.

Vaccines—from Cinderella sector to high value intervention?

Vaccines are one of the oldest biomedical innovations, dating from the late 18th century. Several attempts were made to inoculate against the prevalent and deadly smallpox infection: what gained Edward Jenner the title 'the father of immunology' was that he challenged an eight-year-old boy, after vaccination, with live virus!

For many decades, vaccines for the diseases of childhood—measles, mumps, rubella, whooping cough, etc—dominated the vaccines market. They were typically purchased by governments and health agencies in a bulk manner, with low prices being accepted by manufacturers in light of the high volumes. However, because of these narrow margins, one by one these manufacturers exited the market, in favour of pharmaceuticals in which much higher margins were available. Between 1967 and 2002, the number of producers of paediatric vaccines for the US market reduced from 26 to 12, only five of which were core suppliers. This left a single supplier for eight of the recommended vaccines[29].

The arrival in the last 20 years of some more specialist vaccines for high-risk populations has revived interest in the market. The development and widespread use of HPV vaccines for the prevention of cervical cancer, and vaccines for meningitis are two examples of recently launched products that promise to have a very positive and cost-effective impact on population health.

DNA and cancer vaccines have also stimulated new entrants, and the threat of Ebola, as we saw earlier, generated considerable and urgent vaccine development efforts. In the case of the latter, however, the impact of public health interventions in controlling the disease has meant that only very small high-risk populations have remained for viable clinical trials. Novel trial strategies will be needed to continue the efforts to develop a vaccine for future outbreaks.

So vaccines are enjoying something of a resurgence in commercial interest.

Lesson from the history of vaccines

Incentives

Vaccines make a huge public health impact by controlling the spread of infectious disease. However, sufficient economic incentive has to exist to secure life sciences investment, given the high risk and long time-scales. Sometimes policy-makers can forget this simple truth. (This story is being repeated in some parts of the generic drugs market, where margins have dropped to non-economic levels.) *Sheer volume without margin will not incentivize suppliers.*

Diagnostics—from commodity to crucial targeting technology?

The overall diagnostics story has some parallels to that of vaccines. Despite the undoubted contribution of in vitro diagnostics to health—they are involved in over 80% of medical decisions—they are widely regarded and priced as commodities. Routine 'blood work' is performed at less than $100 for 20 or more tests.

The contribution of high levels of automation and hands-free diagnostic factories, themselves of course welcome innovations, has continued the process of commoditization. Laboratory directors, typically with budgets completely dissociated from treatment costs, are measured by the service levels delivered by these factories rather than their overall contribution to cost-effective care.

The concept of precision medicine, which we will see is a major foundation of the future, is totally dependent on new generations of diagnostics. These will be more complex and costly and ways have to be found to finance their development and to reimburse them adequately, or they will not be available at commercial scale and reliability.

A good example is AMR (see 'Lessons from antimicrobial resistance'). The technology has existed for some time to genotype micro-organisms more rapidly, but the incentives have not been there to create tests with sufficient sensitivity, selectivity, and user-friendliness. This is now beginning to change.

There is an interesting exception to the low relative value accorded to diagnostics, in the area of the screening of donated blood for viruses. Once it was discovered that HIV and the hepatitis viruses could be transmitted through blood transfusions and plasma products, it was quickly demanded that sensitive tests be developed and applied to every unit of donated blood. By that point, 20 000 haemophilia sufferers had died in the USA alone, and legal cases there and in other countries were mounting. The tests, when first introduced, cost between $25 and $35 per unit, out of a 'retail' cost of a unit of blood of $150–200[30].

The FDA now mandates tests for HIV-1, HIV-1/2, Hep B, Hep C, and syphilis; the American Red Cross also tests for human T-cell lymphotropic virus (HTLV) and West Nile Virus. This is despite the fact that the risk of transmitted infection (in the USA) is between 1 in 100 000 and 200 000[31].

In many cases, samples are pooled to reduce the cost per unit; nevertheless the cost is significant. However, the risk of legal challenge that would follow from transfusion of contaminated blood makes the diagnostic seem affordable. In the case of one developing country laboratory (Malawi), the cost of screening donated blood for viruses took up more than half of the hospital laboratory budget![32]

In most segments of the diagnostic market, however, the situation is parallel to that of vaccines: insufficient incentives exist to stimulate and sustain investment in complex development programmes.

A second lesson is that economic evaluation and pricing need to be geared to the overall impact of a product on the care pathway and total treatment cost-effectiveness. Precision medicine will be the test of whether we are able to design these new economic structures.

EGFR (epidermal growth factor) testing in oncology is an interesting case study. Mutations in EGFR have major implications for correct therapy choice. A small panel of DNA tests suffice to identify the common mutations, and—around the same time that the first EGFR inhibitors appeared—commercial tests became available. However, changes in US pricing arrangements (from 'stacked codes' to CMS-defined prices) reduced the reimbursement from $650 to $150, putting the routine use of this test at risk.

One of the most challenging aspects of the genomic diagnostic field is that hospital laboratories can, in most jurisdictions, compete with commercial tests with their own 'home brew' versions. While it would not be practical to outlaw such tests, it is important to ensure they meet the same standards of sensitivity, specificity, and reliability, for the sake of the patients whose treatment is guided by the results.

In principle, as we develop 'precision medicine' (of which much more later) there should be growing strategic synergies between the pharmaceutical and diagnostics businesses. However only one major company, Roche, has well-developed businesses in both. Chiron sold its diagnostic business to Bayer, who had the two businesses for a period, before selling the diagnostics business on to Siemens. Abbott, which once had both, split them into two companies when it spun off Abbvie. Hence, it would appear that the very different economics and technology base of the two make strategic synergies unpersuasive to managements and investors. It will be interesting to see if it is Roche who is right: they routinely involve their diagnostics business in the development of the increasingly targeted drugs in the pipeline.

Lessons from diagnostics

Economics

Allowing the commoditization of a sector carries risks of reducing the willingness of companies to invest, despite the promise of large volumes.

Reimbursement

For diagnostics to play their full part in precision medicine, ways have to be found to reflect their contribution to overall pathway costs in their pricing.

Devices—sustainable or needing drastic cost reduction?

Medical devices cover a very large waterfront, so it is more instructive to look at specific classes of device for our lessons. One of the most successful classes is cardiac electrophysiology, from simple pacemakers to atrial defibrillators.

Given the very stringent regulatory requirements surrounding devices that can either sustain the wearers' lives or cause their deaths, the clinical testing and manufacturing rigour is formidable. As a result, the most sophisticated devices from US manufacturers cost $50 000 and up.

The idea of an implantable cardioverter-defibrillator (ICD) is simple but brilliant. Implant a device that detects arrhythmias—a potent cause of death or heart attack—and restores the heart to its normal rhythm. The battery-powered device is programmable, is implanted in the upper left chest, and a wire leads to the left ventricle via a vein to the heart.

The concept was developed as long ago as 1969 at the Sinai Hospital in Baltimore[33]; however, it was 11 years before the first patient was treated. Much of the development was focused on reliable detection of the onset of ventricular fibrillation and on the ability to deliver an electrical shock to restore a normal rhythm within 15–20 seconds. The concept faced significant opposition within the cardiology profession and little financial backing from grant-giving bodies. Eventually, in 1999, a randomized clinical trial in 1016 people demonstrated the superiority of the ICD over a regimen of anti-arrhythmia drugs (the former had 80 deaths, against 120 with the drugs).

Nevertheless, the conclusion of UK health economists in 2006 was that, despite saving lives, ICDs were not cost-effective, largely as a result of the cost of post-implantation hospitalization[34]. Eventually, following representation by the UK industry group on behalf of five manufacturers, NICE guidance was revised in 2014 to recommend ICDs as a 'possible treatment for people who have had a serious ventricular arrhythmia'. So the total time from conception to positive recommendation in the UK was 44 years (Figure 4.5).

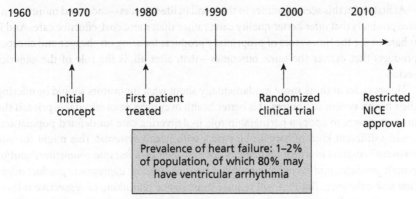

| 1960 | 1970 | 1980 | 1990 | 2000 | 2010 |

Initial concept

First patient treated

Randomized clinical trial

Restricted NICE approval

Prevalence of heart failure: 1–2% of population, of which 80% may have ventricular arrhythmia

Fig. 4.5 Implantable cardioverter-defibrillator—timeline of key events.

The lessons from the ICD case

Development times

Design of complex implantable devices is non-trivial and therefore very time consuming.

Clinical uptake

Even with strong championship from investors, the clinical 'system' is often difficult to shift from conventional practice.

Reimbursement

Clinical trials showing clear efficacy superiority (in this case a demonstrable survival advantage) is a sine qua non, but does not guarantee reimbursement. The latter often requires multiple studies and many years, especially when the initial device is costly. *Coverage with evidence development (reimbursement while value data is collected) could avoid such delays.*

Lessons from elsewhere

We now pose the question: do other industrial sectors hold innovation lessons that life sciences can adopt?

The **aerospace industry**, for example, shares life sciences' high investment levels and long time-scales. They also address markets that have become steadily more cost-conscious.

The first of these characteristics, which amount to very high long-term investment bets, have led to the concept of 'launch customers'—airlines that guarantee initial demand and provide strong customer feedback through development. *Could major payers be persuaded to become launch customers for life science innovations*, prepared to trade their commitment to evaluate and adopt an innovation for price concessions?

The second characteristic, steadily more price-conscious markets, have resulted in aerospace companies designing products that offer more and more energy efficiency and passenger capacity, which amount to lower and lower cost per passenger mile.

At first sight, this seems counter to the trend in life sciences—more and more expensive products that offer better quality care, rather than more cost-effective care. And it is hard to see the innovators of proprietary products focusing on cheaper and cheaper products that deliver the same outcomes—that, after all, is the role of the generics sector.

However, let us think more fundamentally about what innovators should be offering their health system customers. It is better health over time, at an affordable price. If the innovator were to adopt a partnership role in delivering care for defined populations could a different kind of bargain be struck with health systems? This might include *an overall contract in which they supply not only their own but also competitors' and/or generic products, and taking responsibility for such issues as appropriate product adoption and adherence.* But that will require some major rethinking of respective roles—an issue we shall return to in Chapter 8.

The second industry sector to which we might reach for ideas is the *consumer products and retailing* arena. While in principle consumers should need a similar set of products to meet their daily nutritional and household needs, we only have to glance across at someone else's supermarket trolley to see how differently we choose to meet those needs!

The industry responds to this reality in a number of ways. First, it offers a range of products with different levels of quality (real or perceived) at different price points. The second is that is becoming very skilled at identifying individual patterns of purchasing and then promoting, (usually through discounts and 'special offers') those products they know will appeal to the individual shopper.

There are some possible lessons here. The pharmaceutical industry grapples with the fact that affordability of its products varies widely between countries and even within large, diverse economies such as India and China. They are therefore beginning to see the value of different brands at different price points, for essentially the same therapy. These lessons could be extended to developed markets, especially where a product has different indications in which its value varies substantially. Different branding (perhaps supported by different dosing regimens, product support, etc) could enable them to *correctly position what are effectively different innovations carrying quite different value.*

The concept of greater personalization of medicine will be one of the major themes of Chapter 5. But suffice it to say here that the more that innovative industry distinguishes and addresses personal needs and becomes a trusted partner to the patient, the closer it becomes to long-term sustainability.

One of the most innovative sectors over the last few decades has obviously been *personal computer and computing devices.* When the basic PC had seemed caught in an innovative rut of simply faster and faster processing speeds and larger and larger disc drive capacities, manufacturers reinvented themselves as mobile device designers. We have seen a cascade of personal computing devices (tablets, smart phones, and even watches) that have competed on the basis of small size, better and better screens, and more elegant design, with new versions appearing annually.

However, the most important dimension of competition has been software-based, with an ever-increasing number of mobile apps available to meet an ever-widening set of human needs 'on the run'.

Lessons for life sciences? This dynamic sector is becoming more and more interested in healthcare, as the most basic of human needs. Wise life sciences companies will want to partner: work out where products can be combined, resources shared, or joint customer relationships forged to better meet the needs of patients (and also the health-conscious).

While there have been small experiments in this direction, such as 'reminder' apps for pill-taking and prescription-filling, they have only scratched the surface. We shall return to this fruitful idea when we discuss digital health in Chapter 9.

As we reach for solutions for the gaps in translation, the lessons from both life sciences case studies and the experience of other industries begin to highlight both accelerators and inhibitors of innovation:

Accelerators include:

♦ A sense of urgent crisis, as we saw in Ebola and H1N1, which galvanizes companies to act and regulators to challenge their norms. *The challenge is to capture and harness that sense of urgency in tackling diseases that are chronic and serious rather than acute.*

♦ A sound basis of understanding of disease mechanism and the pathways in which the target participates, as was clear for HIV, HCV, Gleevec, and Ivercaptor. In several of our case studies (ulcers, hypertension), *multiple levels of mechanism were probed before today's product portfolio emerged.*

♦ Full engagement of patients and the public, as we saw for HIV, cystic fibrosis, myeloma, and monogenic diabetes. *Active involvement and informed pressure from patients is one of the most potent catalysts we have in changing the innovation process.*

♦ Regulatory flexibility. When regulators are convinced they have a uniquely useful technology on their hands, they often show real flexibility, a sense of partnership with the innovator and a willingness to collaborate internationally.

♦ Collaborative structures, both between the public and private sectors and between companies, can achieve what no single organization can.

♦ Pricing and reimbursement problem solving. In the face of economic pressure, the failure to engage effectively with this issue can be a show-shopper. The value impact on the total care pathway can often be a major factor, as we saw in RA.

♦ Precision medicine. Therapeutic breakthroughs are often based on precise application to the patients most likely to benefit, along with the diagnostic that makes that possible (as we saw in cystic fibrosis and monogenic diabetes).

Inhibitors can be both scientific and societal:

♦ A failure to invest sufficiently in uncovering the mechanism of disease and developing in vitro or reliable animal models for testing treatments.

♦ Duplicative, rather than collaborative research where basic understanding is missing or costs of development are high, resulting in higher-than-necessary upfront investment.

♦ Unproductive tension and mutual finger-pointing between the industry producing products and the systems that need to buy them.

- A lack of attention to the basic economics that drive innovative companies to invest, and a lack of adaptive reimbursement models where high upfront costs result in higher long-term savings.
- Clinical conservatism, pure and simple.

Summary of Chapter 4

There are many points of light that show the way forward to more productive and sustainable medical innovation. A sense of urgency, strong patient and public engagement, regulatory flexibility, and collaboration models need to characterize our solutions, as we move into an era of precision medicine. But old models die hard, especially when the companies and agencies that use them have seen great past success.

In Chapter 5 we build on these lessons to lay out the most important steps to make effective translation the norm rather than the exception.

References

1. Wikipedia page, http://en.wikipedia.org/wiki/Management_of_HIV/AIDS and references within it, accessed 01 Sep. 2015.

2. Centers for Disease Control and Prevention, http://www.cdc.gov/h1n1flu/information_h1n1_virus_qa.htm, accessed 01 Sep. 2015

3. European Strategy for Influenza A/H1N1 Vaccine Benefit-Risk Monitoring published by the European Medicines Agency, Oct 2009 http://www.ema.europa.eu/docs/en_GB/document_library/Report/2010/01/WC500044933.pdf, accessed 01 Sep. 2015

4. Phillips, L.D. et al., Modelling the risk-benefit impact of H1N1 influenza vaccines. *The European Journal of Public Health*, 23(4):674–678, http://eurpub.oxfordjournals.org/content/23/4/674, accessed 01 Sep. 2015

5. WHO press briefing by Dr Marie-Paule Kieny, 19 November 2009; transcript available http://www.who.int/mediacentre/vpc_transcript_19_november_09_kieny.pdf, accessed 01 Sep. 2015

6. **Vitelli, A. et al.**, Potential "universal" vaccine tested at FDA, cited at http://www.fda.gov/BiologicsBloodVaccines/ScienceResearch/ucm353397.htm, accessed 01 Sep. 2015

7. Recommendations for Accelerating the Development of Ebola Vaccines, Feb 2015, published by the University of Minnesota, the Center for Infectious Disease Research and Policy, in conjunction with the Wellcome Trust http://www.cidrap.umn.edu/sites/default/files/public/downloads/wellcome_trust-cidrap_ebola_vaccine_team_b_interim_report-final.pdf, accessed 01 Sep. 2015

8. Unprecedented Microbial Diversity Found in Amazonian Tribe, blog by Julia Paoli, Scitable, published by *Nature Education*, April 22, 2015 http://www.nature.com/scitable/blog/viruses101/unprecedented_microbial_diversity_found_in, accessed 01 Sep. 2015

9. **Hawkey PM, Jones AM.** 2009. The changing epidemiology of resistance. *J Antimicrob Chemother*, 64(Suppl. 1):i3–10.

10. **Liu B, Pop M.** 2009. ARDB—Antibiotic resistance genes database. *Nucleic Acids Res*, 37:D443.

11. **Davies S, Grant J, Catchpole M.** 2013. *The Drugs Don't Work: A Global Threat*. London: Penguin.

12. **Jim O'Neill,** Tackling drug-resistant infections globally, Final Report, May 2016, available at amr-review.org http://amr-review.org, accessed 01 Sep. 2015

13. Product Characteristics for Kalydeco, published by the European Medicines Agency, available at http://www.ema.europa.eu/docs/en_GB/document_library/EPAR_-_Product_Information/human/002494/WC500130696.pdf, accessed 01 Sep. 2015

14. A study to confirm the long-term safety and effectiveness of Kalydeco in patients with cystic fibrosis, details available at https://clinicaltrials.gov/ct2/show/NCT02722057, accessed 01 Sep. 2015

15. FDA FY2012 Innovative Drug Approvals, accessed at www.fda.gov/AboutFDA/ReportsManualsForms/Reports/ucm276385.htm, accessed 01 Sep. 2015

16. **John Fauber,** Cystic fibrosis: Charity and industry partner for profit, http://www.medpage-today.com/Pulmonology/CysticFibrosis/39217, accessed 01 Sep. 2015

17. **Houghton M.** 2009. Discovery of the hepatitis C virus. *Liver Int*, 29 Suppl 1:82–88.

18. **Johnson JO, Mandrioli J, Benatar M,** et al. 2010. Exome sequencing reveals VCP mutations as a cause of familial ALS. *Neuron*, 68:857–64.

19. **Ian McDonald,** Dementia risk reduction strategies prove beneficial, March 13, 2015 http://dementiaresearchfoundation.org.au/tags/finger-study, accessed 01 Sep. 2015

20. **Mestre-Ferrandiz J, Sussex J, Towse A.** 2012. The R&D Cost of a New Medicine. Office of Health Economics, London. https://www.ohe.org/publications/rd-cost-

21. Guidance for industry, Developing drugs for the treatment of early stage disease, FDA guidance document, February 2013 http://www.fda.gov/downloads/drugs/guidancecomplianceregulatoryinformation/guidances/ucm338287.pdf, accessed 01 Sep. 2015

22. Dementia Consortium website http://www.dementiaconsortium.org, accessed 01 Sep. 2015

23. **Lachmann PJ.** 2012. The penumbra of thalidomide, the litigation culture and the licensing of pharmaceuticals. *QJM*, 105:1179–89.

24. Bortezomib monotherapy for relapsed multiple myeloma, National Institute for Health and Social Care Excellence, Technology Appraisal Guidance, 23 Oct. 2007 http://www.nice.org.uk/guidance/ta129/documents/department-of-health-summary-of-responder-scheme2, accessed 01 Sep. 2015

25. **Coulton, L.** et al. Risk-sharing schemes worldwide, http://www.ispor.org/research_pdfs/35/pdffiles/PHP15.pdf, accessed 01 Sep. 2015

26. **Nowell, PC.** 2007. Discovery of the Philadelphia chromosome: a personal perspective. *J Clin Invest*, 117:2033–5.

27. **Rowley, JD.** 1973. A new consistent chromosomal abnormality in CML. *Nature* 243:290–3.

28. **Godbehere P, Wareing P.** 2014. Hypertension assessment and management: role for digital medicine. *J Clin Hypertension*, 16:235.

29. **Danzon PM, Sousa Pereira N.** 2011. Vaccine supply: effects of regulation and competition. *Int J Econ Bus*, 18:239–71.

30. **Starr D.** 1998. *Blood: An Epic History of Medicine and Commerce*, New York: Knopf.

31. Tissue Safety & Availability, FDA regulations cited http://www.fda.gov/BiologicsBloodVaccines/SafetyAvailability/TissueSafety/ucm151757.html, accessed 01 Sep. 2015

32. **Lara AM, Kandulu J, Chisuwo C, Kashoti A, Mundy C, Bates I.** 2007. Laboratory costs of a hospital-based blood transfusion service in Malawi. *J Clin Pathol*, 60:1117–20.

33. **Mirowski M, Reid PR, Mower, MM,** et al. 1980. Termination of malignant ventricular arrhythmias with an implanted automatic defibrillator in human beings. *New Engl J Med*, 303:322–4.

34. **Buxton M, Caine N, Chase D,** et al. 2006. A review of the evidence on the effects and costs of implantable cardioverter defibrillator therapy in different patient groups, and modelling of cost-effectiveness and cost-utility for these groups in a UK context. *Health Technol Assess*, 10:iii–iv, ix–xi, 1–164.

Chapter 5

Seven steps to sustainability based on precision medicine

Unless we change historical trends, medical innovation is unsustainable. Bioscience advances will not be translated into patient benefit as they should. The potential of much of the exciting advances in genomics and other cutting edge technologies described in Chapter 2 will go unrealized, or realized only after long delays and huge investments. We saw in Chapter 3 that the problems underlying declining innovation productivity are many and varied. There are gaps in translation at every stage from discovery, through development, regulatory approval, reimbursement, and adoption to ultimate patient benefit. For too many innovations, this means a journey of more than 20 years.

However, in Chapter 4 we saw several examples of rapid translation, regulatory flexibility, and speedy adoption that encourages us to believe that reform of this process is possible. Our challenge is to make this flexibility, speed, and uptake the rule, rather than the exception.

However, closing the innovation gap will be far from straightforward. It qualifies as a 'wicked problem' using Patricia Seeman's definition[1]. Solving these kinds of problems involves disruption to customary approaches, the need for speed and agility, and convergence of factors and actors. In such situations, she argues, it is not merely a question of thinking the problem through and implementing the solution. We are dealing with a dynamic system, involving many stakeholders, and the best solutions may be developed at the coal face on specifics, rather than abstractly, in an ivory policy-making tower. All of this seems to fit the medical innovation translation problem!

Most of all, we need some significant *mindset changes* from most, if not all, the participants. Patients and payers need to be involved from the start, not just at the end. Academics need to see themselves as part of an innovation process, not just independent researchers. Regulators, instead of making 'final' approval decisions about new products at one point in time, need to accept greater uncertainty from formal trials and adapt and adjust in light of real experience. The focus of companies and health systems should also be on working together to ensure the right treatment reaches the right patient at the right time, rather than assuming a 'free-for-all' once marketing approval is granted. We need to be prepared to experiment with new payment structures. Throughout, we need to strive for joint decision-making through the process, rather than a sequence of disconnected decisions from separate agencies.

Building on principles such as these, in this chapter we explore how the new and more sustainable biomedical innovation process should work. But before describing these changes, we should examine their scientific basis, as some of the scientific advances reviewed in Chapter 2 provide new tools for this new process. Without them we could not make many of the changes needed.

Build on new science and technology

Science has begun to deliver important new tools to help address our five (T0–T4) gaps in translation. There has been progress on at least four relevant fronts: basic biology (T0 and T1), development tools and technology (T2), the understanding of human behaviour (T3), and data analytics and machine learning (T4).

Preclinical science—for T0 and T1

We have seen how vital it is to understand disease mechanisms if we are to tackle unmet medical need. Many advances in basic biology inform this challenge of finding disease-specific research avenues, and the T1 challenge of translating this research into candidate products.

Our greater understanding of the role of genetics and epigenetics, and of the role of the complex networks of systems biology in the molecular basis of disease, are opening up more precise and intelligent interventions. This is especially true where the genetic or phenotypic basis of disease has emerged with a new clarity and simplicity. We are also assembling new preclinical tools. Organ simulation software, stem cell-derived in vitro tests of efficacy and toxicology, and more refined animal models of disease can perform better screening of candidate products before they go into the clinic. Then, experimental medicine techniques such as micro-dosing and PET imaging are enabling us to track how potential therapies behave in the body.

There have been some stunning successes in moving from basic biological discoveries to starting points for potential treatments. The examples of juvenile diabetes and cystic fibrosis in Chapter 4 show the power of such insights. Sometimes, however, even when we know fairly precisely what the flaw is in a patient's genetic and hence metabolic make-up, the treatment designed around it is only partially successful: cellular pathways adjust to the treatment to defeat it.

As a current example of a T1 gap, take Duchenne's muscular dystrophy. We know a great deal about the genetics of the disease and treatments based on the exon-skipping approach have a positive initial impact—measured by the 5-minute walk test—but the effects diminish over time. At the time of writing, it isn't clear why. Similar phenomena occur with many cancer treatments. Even Gleevec, we saw, progressively induces resistance, forcing the development of similarly targeted drugs that can evade this resistance.

When we come to genetically and metabolically more complex diseases like the neurogenerative disorders (Alzheimer's, Parkinson's, ALS, etc) the 'single target' drug discovery philosophy so popular in recent years seems to break down completely. We are beginning to realize that many of the drugs discovered via either serendipity or screening against animal models of disease have earned usefulness *because* they do not have a single site of action.

For complex diseases, we need to move from the analytical, 'drill-down' approach (single gene-to-protein-to-drug) that has been so appealing in the post-genomic era. We need a synthetic 'build-up' approach in which we map all the pathways of significance and investigate the effects of different types of interventions on the overall pattern of gene expression, protein activity, and metabolic balance. This is 'network pharmacology'.

Genomics is just one of the 'omics' now at our disposal. We use a variety of tools to accurately map and track as many facets of the patient's biological status quo as we have tools to do so. A fascinating recent study did just this for one individual in their 50s. This study tracked the multi-omic parameters of this single individual (in fact, the lead investigator, Mike Snyder) for some weeks, through two infections (rhinovirus and respiratory syncytial virus). It demonstrated how the second, more serious infection triggered latent Type 2 diabetes, for which the individual was inherently prone[2]. The layers of information in this study—genes expressed, proteins manufactured, cytokines circulating, etc—have yet to be fully mined, but we can already see how this molecular 'X-ray' approach will unearth links and networks never before guessed at.

What is also clear is that we will be able to carry out this kind of molecular profiling—involving genomics, epigenomics, proteomics—reasonably affordably, perhaps for around $1000 per individual. So we will need the ability to analyse and combine these complex streams of data on a patient-by-patient basis to gain an overall picture and time course of the disease and its impact on the patient. This is one of the challenges we will return to later.

We are probably some years away from a full understanding how these networks become dysfunctional in some specific diseases and how therapies can re-balance them. But increasingly we can ask and answer the 'five whys' about diseases that still resist our ability to intervene successfully. We see death of neurons: why? Cell death, apoptosis, is being triggered by the protein Bcl-2: why? The level of Bcl-2 is higher in this disease than in health: why? And so on.

To be provocative, it may be that some Asian scientists adapt more readily to this holistic thinking than some of their Western counterparts. The Western mind is much more focused on the object in the foreground, whilst Eastern thought patterns seek the holistic picture[3]. I am sure this assertion will generate plenty of criticism from my Western colleagues!

Once a drug is found that perturbs biological circuitry in a promising manner, better biomarker-based tests are coming available for how it will perform in the body. Both efficacy and safety are beginning to be predicted in this way. Isolated iPS-based cardiomyocytes can be used to test cardiac response for a range of different patient genotypes. Similar experiments are being carried out on neurons, in vitro.

Stem cell-derived hepatocytes—the liver cells most critical for drug metabolism and potential liver toxicity—have proved to be a more challenging model system, but in time these too will be available. Using cultured human hepatocytes we are already able to create a 'liver on a chip', and these whole organ test systems will grow in importance. And, with support from the US Department of Defence (for toxin testing) researchers at Harvard's Wyss Institute are already assembling a 'human body on a chip' with cells from 12 different organ types[4]. All of this will reduce the relevance of animal models, long the backbone of pre-clinical testing.

'In silico' biology is also making an impact. We have had relatively crude computer models of the heart for some time to enable us to anticipate the overall impact of a drug or device on an organ in vivo. This type of modelling is now extending to finer detail: predicting the behaviour of an individual cell. There is now even a journal devoted to this new 'in silico biology' discipline[5].

Science and technology for clinical development—T2

A second group of advances—in clinical research design and electronic monitoring— have great potential to improve the productivity of clinical development, the T2 challenge. Creativity here has been constrained by real or perceived regulatory barriers, but adaptive trial designs and biomarker-stratified patient populations are beginning to have a substantial impact.

Biomarkers derived from the 'omic' science just described are increasingly being used not just to identify targets for drug discovery but also for selecting patients for clinical trials. This approach is in active use in Alzheimer's[6], where both imaging and cerebrospinal fluid markers are in use, in ALS,[7] and of course in cancer[8]. It is now almost unthinkable to run a clinical trial for a new cancer medicine without selecting patients on the basis of their patterns of mutations or the receptors expressed on their cell surface.

The better targeted a trial is, the higher the responses and the more likely regulators are to consider early approvals, as we will explore later.

Biomarkers are also gaining attention as surrogate markers of treatment response, making a reality of the 'therametrics' concept that Bill Rutter and I worked on at Chiron nearly 20 years ago. This will be of particular importance in diseases such as Alzheimer's in which clinical measures of treatment response may take many years to establish.

The wider availability of electronic medical records can assist in scaling trials and finding subjects, and electronic tools are increasingly in use for recording results in the course of trials[9]. The use of remote electronic monitoring of clinical trials can replace or substantially reduce the need for human monitors, but more importantly to identify missing or erroneous data entries in real time[10]. It can also help the important process of sharing those results with the patients.

Behavioural sciences for T3

Ensuring effective diffusion and adherence is the T3 challenge. Here it will be the social and behavioural sciences that offer the best insights. The diffusion of innovation is beginning to be understood as a community or network problem, influenced by practitioner psychology and connectivity as much as it is by hard-edged clinical evidence. New light is being shed on the long-lamented problem of poor patient adherence to treatment by psychological profiling of patient beliefs, needs, and concerns and their effects on behaviour.

For too long, the spread and appropriate use of innovations have been left to a combination of commercial promotion and chance encounters between doctors and information, via journals, conferences, and conversations.

Now the application of insights about how and why individuals make decisions is beginning to create the basis for a more engineered approach. What are some of the insights recent research has brought us? Firstly, that what we regard as rational decisions by rational actors is often far from that. In most situations, we tend to avoid risk rather than seek benefit. In fact, we do not independently assess the risks and benefits of a course of action: we tend to combine the two. In other words, an action we think will be beneficial we assess as less risky.

We might expect regulatory assessors, whose job it is to assess both risk and benefit for a new medicine, to be free of such a 'heuristic' (rule of thumb), but they are not. They too connect perceived benefit with perceived risk[11]. However, interestingly and perhaps surprisingly, this study on staff at European regulatory agencies also explodes the popular perception that regulatory assessors are inherently risk-averse. Most are risk neutral. Their attitudes to risk are the same complex mixture of personality and situation as is true for the rest of us.

One striking finding is that the 'conscientiousness' trait among assessors is a strong predictor of the assessment of benefit. Conscientiousness is one of the 'Big Five' personality traits (extraversion, openness, neuroticism, conscientiousness, and agreeableness) that are used to analyse personality[12]. This means that the most conscientious assessors are more likely to seek and see benefit in a new product.

Research shows the network effect to be particularly influential with doctors, who consistently listen to other doctors and their experience more than any other sources. Setting up the right reference networks is therefore crucial to securing adoption. Pharmaceutical companies have long understood the power of 'key opinion leaders'. However, in the Internet age, the voices of these key opinion leaders may be increasingly drowned out by the much larger number of 'doctors like me' sharing their views on medico-social networks.

We are also learning more about patients' thought processes as they make the decision of whether and when to fill and comply with a prescription. Patients commonly harbour wrong perceptions about an innovation, its impact on their lives, and the risks it might bring. Risk is a particular concern, given the tendency to seek to minimize risk rather than to maximize benefit.

Network psychology is also important for patients. Doctors' recommendations are valued and they are trusted by most patients, but the latter often take more notice of 'people like them'—colleagues, friends, family members, and even casual contacts, physical or over the Web—who seem to face the same health issues and choices.

Information technology for T4

The fourth gap to close involves developing effective feedback from real world clinical experience and outcomes for the better use of existing products and for clues to the development of new ones. This is an information management challenge. We are finally entering the era of widespread and comprehensive electronic medical records.

The resulting database, along with health-related social media exchanges, can be mined using artificial intelligence approaches. This promises totally new insights into how conditions and their therapies affect patients, particularly in the case of multiple morbidities and treatments, which are never tested in clinical trials.

As a result of all these data streams we are also in the era of 'Big Health Data'. This is an over-hyped and under-analysed slogan, born of the tendency of IT practitioners to announce major revolutions (remote computing, cloud-based solutions, on-demand computing, and the like). However, large-scale data—if it is broadly accurate—is able to separate the signal from the noise on key clinical questions, such as the link between blood pressure and cardiac risk.

Fundamentally, it is not the sheer scale of the data itself that promises the most, but also the ability to deduce unexpected insights from within it. In other words, it is not our storage and processing capacity for terabytes of health data that matters, it is the quality of comparable data, the ability to make and test correlations, and to focus on specific insights that are most likely to be relevant to small groups of patients.

If we are to distil the most we can from data analytics, we need to create data commons. This need not imply placing all the data in a giant database, with its practical and ethical challenges: standard formats, federated databases, and remote querying will achieve the same goal. Health data, of course, comes in various types and modes, historically in incompatible forms and stored in systems that do not intercommunicate. They include genomic and other 'omic' data, the results from clinical trials, medical images of various kinds, electronic medical records, patient-reported outcomes, and, increasingly, data captured in social media. The use of artificial intelligence can begin to bridge these data sets, as long as reliable patient identification can be assured.

We are just at the beginning of applying artificial intelligence to this task, and we will return to this in Chapter 5.

In summary, as we redesign the innovation process, we can build in science and technology that did not exist when the current process was created. All the gaps in translation can benefit, from initial discovery through to learning from real world experience. We now turn to how we can apply both new technology and a new mindset to reinvigorating the medical innovation process.

Seven steps to sustainable medical innovation based on precision medicine

The worsening productivity of life science innovation over time represents a threat to its very sustainability. We need radical changes—changes that help bridge the multiple 'gaps in translation'. I propose seven steps to sustainability that can transform the process. *Each represents a fundamental change of mindset* across most of its stakeholders. Some of these changes in mindset are more advanced than others—all challenge the status quo of the last 20–30 years. They are all actionable: I have personally been involved in developing or implementing many of them. Together they build the precision medicine of the future.

The seven changes of mindset and of practice are:

1. **Advance the molecular definition of disease and the application of systems biology.** We need a more decisive move from a classic definition of diseases—in terms of the symptoms they cause and/or where in the body they appear—to a definition in terms of aberrant, defective, or unbalanced molecular mechanisms at the cellular level. And we need to marry this with a recognition that singular target-based innovation rarely works: we need a systems biology approach.

2. **Partner academia and industry in more collaborative, impact-oriented research.** We need to extend the 'open innovation' approach in which academia and companies invest together and share IP. We need to define new pre- or non-competitive spaces, especially in work on disease mechanisms and disease models. And we need to provide for new types of links and incentives to break down the barriers between these two worlds.

3. **Move decisively to a more adaptive approach to development, trial and approval design.** We need to build on successful experiments in more flexible trial design, development pathways, and regulatory appraisal to a globally accepted adaptive approach. This involves collaborative design of the evidence package needed to secure approval and reimbursement, and greater teamwork through the process.

4. **Create new reward and financing vehicles for leading edge innovation.** We need to move from reward systems based purely on unit sales of products, irrespective of outcome, to rewarding innovators for positive outcomes, patient by patient. We also need to design financing mechanisms that bridge between cost-effectiveness and affordability. We must be able to accommodate high-cost precision therapies that offer cures and so generate long-term returns for the system.

5. **Engineer tools and systems for faster and better innovation adoption and adherence.** We need to move from reliance solely on promotion to doctors and passive patient participation to a disciplined approach to establishing new pathways of care. These will be based on modern behavioural science, clinical decision support, and other digital technologies.

6. **Develop an infrastructure for real-world data-driven learning.** We now have the opportunity to study in large populations how lifestyle and treatment choices lead to outcomes, learning from every patient as if in a clinical trial. New analytical tools will empower this.

7. **Bring patients into the mainstream of decision-making and engage them wholeheartedly throughout the process.** It is time to move from a process and mindset in which patients are regarded as passive subjects for clinical trials and recipients of products and procedures. Their input and engagement needs to be sought along the whole innovation chain: on treatment benefits, acceptable risks, optimal clinical trial design, adherence support, and outcomes.

As a result of all these changes we can build a new *precision medicine future*. We will have the tools to practice more personalized, precise medicine at every stage of the innovation process. This is the direction of travel enabled by the seven steps. Together they have the power to tackle the five gaps in translation: Steps 1 and 2 are designed to address the research stage gaps T0 and T1; steps 3 and 4 will overcome gap T2 and so improve the productivity of development and the rate of commercial success; step 5 will lay the basis for dealing with the adoption and adherence issues in T3; and step 6 will make real world health data a growing source of insight for the whole R&D enterprise, closing T4; step 7 brings patients more fully into the enterprise, bringing value at every stage.

Let us now describe in more detail what these new mindsets would lead to, in terms of practical change. If we look back at Chapter 4, we will see we are building on successful examples from the past.

Advance the molecular definition of disease and the application of systems biology

The classic definition of diseases has been in terms of the symptoms they cause and/or where in the body they appear. This was the best that medicine could do when external observation of the patient was the only or primary means of diagnosing disease. The powerful new tools of molecular biology are reinterpreting disease in terms of aberrant, defective, or unbalanced molecular mechanisms at the cellular, organ, or organism level. Molecular level diagnosis becomes a real possibility.

Such an approach brings effective therapy immediately closer. Molecular diagnostics can separate diseases with similar symptoms but different underlying causes—and often suggest a different starting point for intervention. Accurate genotyping, as we saw in the case of neonatal diabetes and cystic fibrosis, can transform therapy. Likewise, for the more genetically complex diseases, accurate phenotyping can do the same.

In some cases, the reverse will be true: different symptoms may turn out to flow from common molecular causes, often related to the immune system. Cancers that have different organs of origin but similar mutational signatures call for similar treatment.

In yet other cases, new insights into the causative factors for a disease completely overturn how it is classified and treated. The classic example from Chapter 4 was that of stomach ulcers, long regarded and treated as the result of excess stomach acidity before the discovery that it was caused by an infection of the bacterium *H. pylori*.

Our deepening level of molecular insight will ultimately overturn the whole way in which diseases are classified. This will have very broad implications—from research to medical education. But the longer we delay the change, the longer we will research, develop, and treat on an anachronistic basis. The challenge represented by this major shift to molecular taxonomy of disease is now being tackled by a project sponsored by the Innovative Medicines Initiative (IMI) in Europe[13]. A similar project was called for in the USA by a National Research Council committee and a report is now available[14,15] titled 'Toward Precision Medicine', making clear that such a molecular taxonomy is a critical path item en route to true precision medicine. The US report highlighted the way in which lung cancer has been progressively redefined in terms of its mutational pattern and hence drug sensitivity. Until 1987, lung cancer was divided by cell type: adenocarcimoma, squamous, and large cell. Then *KRAS* mutations were identified. By 2004, *EGFR* mutations had been added, and by 5 years later there was a total of eight driver mutations identified[16] . That number continues to rise.

Ultimately, of course, this task of disease redefinition will need to be one for the whole global biomedical research community. If we undertake only regional or even national efforts, we will repeat medical history in which conditions are described differently in different countries, complicating international comparisons of medical data.

Of course, this transformation in the basis of diagnosis also vastly complicates medicine itself. It multiplies the number of diseases a doctor must know and reduces (in most cases) the number of patients that any given therapy fits. This poses a major challenge for clinical science and for clinical education.

Despite a lot of genomic research in areas like diabetes, it may be some time before it makes a decisive impact on major chronic diseases. However, new phenotypic tests are already beginning to do so. In asthma or COPD, it has been conventional to distinguish those with mild symptoms and those with more severe impairment of lung function. However, it turns out that it is much more meaningful to track the presence or absence of eosinophilic airway inflammation. This may be found in patients with milder symptoms but such patients are of greater risk of major exacerbations and death. Drugs to combat airway inflammation should be targeted to those patients, who can even be identified by eosinophils in the blood[17]. These should be the high priority patients, not those identified via the traditional method—FEV (forced expiratory volume) scores.

The process of realigning medicine itself—including its terminology and training—to molecular insights, as opposed to symptoms, is likely to be the rate-limiting step in the practical application of the new biology. However, we can begin to see the potential advantages by looking at those diseases where the process is already well advanced, such as leukaemia. Doctors already recognize more than a dozen sub-types on the basis of pathological examination of the cells affected. These diseases are now in the process of further sub-categorization using genomics and epigenomics, with profound implications for successful treatment.

Once a disease has been pin-pointed in molecular terms, the challenge is to find the right intervention point(s). Drug discoverers must concede that single, disease-specific intervention points rarely exist in biology. The cell is made up of interlinked self-correcting networks. Imbalance in these networks causes disease, but correcting this imbalance is often not straightforward, as either the network quickly adjusts around the treatment, or builds up resistance over time.

For much of the last 20 years, an overly simplistic model drove much of drug discovery. The formula ran: find the gene(s) linked to the disease, isolate and sequence the corresponding protein, determine its structure and that of its active site. Then design a wide range of molecules to fit the site, and screen them to find the tightest binding, most drug-like, and take it into clinical development. The model has appealing directness and lends itself to 'industrialization' via combinatorial chemistry and high throughput screening. It has worked in infectious disease, as we saw in HIV and HCV, and also in those diseases with a clear and straightforward genetic cause. However, for complex, chronic diseases, for the most part it hasn't worked. When we come to genetically and metabolically very complex diseases like the neurogenerative disorders (Alzheimer's, Parkinson's, ALS, etc) the 'single target' drug discovery philosophy breaks down completely.

We have ignored the systems nature of biology, with its interlinked pathways and feedback loops. We have undervalued the development of preclinical disease models and their power to predict human response. We have also underestimated the human factor—the experience and instincts of successful drug discoverers and clinical pharmacologists. These kinds of experience and instincts delivered some of the most valuable of the current medicines, drawing on much less biological knowledge than we have today.

The approach we have taken is best described as reductionist. We now recognize that we also need the 'synthesist' viewpoint—understanding the system and the likely effects of an intervention. Of course, without the work of the reductionist to identify the individual components and their roles in biology, there is nothing for the synthesist to work on. But once the pathways are established and their role within the overall cellular system described, the whole can be modelled mathematically and its response to intervention more accurately predicted. Systems biology is defined by Wikipedia as 'the ability to obtain, integrate and analyse complex data sets from multiple experimental sources using interdisciplinary tools'. These are the approaches we need.

When drug resistance emerges we have to go back to these pathways and deduce where biological evolution is conspiring to defeat us. We saw that Gleevec, the first really targeted oncology drug producing startling initial successes in a range of tumours, progressively induces resistance in the cancer. Drugs with similar mechanisms that can evade the tumour's resistance then have to be developed.

We are beginning to suspect that many of the drugs discovered in the past via either serendipity or screening against animal models of disease have been useful *because* they do not have just a single site of action. We met an apparent example of this in Chapter 4, where multiple mechanisms appear to underlie the action of methatrexate in auto-immune disease, including: inhibition of purine metabolism; inhibition of the activation of T-cells and suppression of intercellular adhesion molecules on their surface; selective down-regulation of B cells; increasing CD95 sensitivity of activated T cells; inhibition of methyltransferase activity; the inhibition of the binding of Interleukin 1 beta to its cell surface receptor[18,19,20].

For those diseases where we lack a real understanding of the pathways involved, we need to fund and coordinate more work along these lines. Both T0 and T1 gaps stand to benefit from more precise molecular disease definition, better disease models, and more systems-oriented approaches. Without them we will see more basic biology that fails to feed through into candidate products.

Partner academia and industry in more collaborative, impact-oriented research

'Impact' is an increasing watchword on the part of public funders and research charities. They want reassurance that the programmes they fund will ultimately translate into treatments. Universities seeking external funding from industry need to show they are alert to this goal also. So the days in which pure academic curiosity and peer review dominate all research funding are now slipping away.

This demand for demonstrable impact has profound implications for academic research. It can be—or be perceived as—a negative force, if it reduces spending on the kind of basic science of genuinely international quality that opens up new avenues of research. Without these, we will see little progress further downstream. World-class scientific talent in top quality research institutes needs to be supported as much as ever. However, many of the best basic scientists are also fully alert to the potential of their discoveries. The kind of retrospective analysis of impact now being performed by funding agencies does not seem to penalize them, but reveal their level of long-term influence.

However, there remains a major T1 gap between such basic advances and practical product candidates. In addressing this, it is important to realize an important difference between most life science discovery research and that in other sectors. Unlike, say, many branches of engineering, the interests, skills, and resources of the academic life science researcher stop well short of product design and development. And there remains a cultural gap with the commercial sector, with a mismatch in success criteria, timescales, and publication strategies. Yet most commercial companies currently do not engage with universities until they see the prospect of a product.

We need close, creative collaboration across this interface, with academics and industrial researchers working together. The language of 'Technology Transfer', with its image of a hand-off from one to the other, is usually inappropriate in life sciences.

Conscious of the gap, and the potential if we can close it, some companies are sponsoring fellowships in which scientists spend time in both environments. Others 'embed' their own researchers in universities. Most innovative companies are seeking to nestle their own labs as closely alongside leading academic centres as possible. A visitor to the Cambridge/Boston cluster will see Amgen, Novartis, Pfizer, Merck, and AstraZeneca among the companies in the neighbourhood. A visitor to Cambridge (UK) will see a huge AstraZeneca site rising alongside the Addenbrookes Hospital, the home of the renowned Laboratory of Molecular Biology with its ten past Nobel prizes.

Innovation 'incubators', 'accelerators', and 'escalators' are springing up alongside most major academic centres. These are vital mechanisms, but need mentors with practical industry experience to help nurture projects at the pre-company stage. Too many spin-outs with too little critical mass will not advance T1 translation very much.

However, it is not only individual company-academic collaboration we need. Many of the toughest problems, such as researching disease mechanisms, finding biomarkers to distinguish them, and developing models of drug response are best tackled in communities. Enabling programmes like these belong in the collaborative 'open innovation' space. The SGC, as described in Chapter 2 is just one model of the kind of consortium approach we need. The Innovative Medicines Initiative, that we will describe in Chapter 6, is another.

Funders that are genuinely interested in translation should support such public–private or private–charity consortia to provide intellectual infrastructure that moves the whole field forward. Examples of worthy targets are unsolved disease mechanisms, better preclinical models of disease (in silico, in vitro, and animal), the development and validation of robust biomarkers, and the creation of databases that capture collective learning in areas like predictive toxicology and reasons for past development failures.

Collaborative models, however, immediately throw up questions of IP, both for universities and for companies. Of course, the seeking and granting of patents have long dominated discovery in biomedical science. This system—excluding others from a patented area to secure competitive advantage, a period of exclusivity and so a more secure return on investment—has been the standard business model and investment dynamo for life sciences. In fact, without patents we would not have had commercial investments in research and the products that deliver for patients today, both prior to and long after patent exclusivity.

The model also fitted comfortably within the 20th century path to success in the pharmaceutical industry—the so-called FIPCO (fully integrated pharmaceutical company) concept. Companies would conduct alone, and in house, all the key steps en route from the biological insight reported in the scientific literature to a marketed therapy.

However, in recent years the patent-driven model has been under fire for a number of reasons. Patents and so-called 'patent thickets' have been seen as inhibitors of innovation, especially where the patent holder is not exploiting it directly, but simply hoping to collect a royalty from more enterprising developers. The very existence of patents has also been a major target for activists seeking to address the health needs of the developing world.

Without IP protection of specific products, of course, companies will not invest in their development. We weaken that protection at our peril, in the absence of an alternative mechanism to deliver a period of exclusivity. However, there is much intellectual infrastructure of the kinds listed above—short of end-product-specific IP—that can and should be developed collaboratively. So the move is now towards an 'open innovation' mindset within a more collaborative R&D model.

Open innovation defined as 'the process of innovating with others for shared risk and reward to produce mutual benefits' has evolved significantly over the last two decades. It is based on the realization that independently developing what are essentially common tools is wasteful. In contrast, open innovation collaborations can vary in terms of number of partners, R&D stages involved, and the level of 'openness' in knowledge and IP sharing. We can also classify such partnerships in terms of their goals, which are summarized in Box 5.1.

So open innovation goals vary widely—creating products, developing tools & models, building information databases, and using the skills of the 'crowd' in problem-solving.

In a study for the Wellcome Trust, CASMI examined a wide range of collaborative arrangements in life sciences and drew out some lessons on how best to make them work[21]. These guidelines (see Box 5.2) focus on clarity of objectives, systems for monitoring progress and clear upfront agreement on IP issues.

It is interesting to note from this review that university technology transfer offices are more often seen as a barrier than a facilitator for these kinds of open innovation, as a result of their frequent focus on capturing maximum value from individual assets, rather than contributing them to a successful collaboration.

To ensure that these problems are properly addressed, professional management of consortia is vital. 'Travelling hopefully' together is not good enough: tough go/no-go decisions often have to be made, or participants persuaded to re-engage when there

Box 5.1 Classes of open innovation

Partnerships focused on single disease or therapy types: Therapeutically focused collaborations often cover multiple R&D stages; broader partnerships typically relate to a specific step, such as compound screening.

Development of tools, standards, and models: Collaborative development of bio-markers, or tools such as cell-based toxicology screening. A number of the projects run by the Critical Path Institute in the USA and IMI in Europe have pursued such 'enabling technologies'.

Determination of protein structures for potential drug targets: This was the initial focus of the SGC, enabling shared private sector and research charity investment in creating generally available starting points for drug discovery.

Sharing genomics insights: This is also a growing area for national and international collaboration. CLARITY, a clinical genetics challenge run by the Boston Children's Hospital, invites clinicians across the world to use genetic data to diagnose cases. The GA4GH (Genomics Alliance for Global Health) is a broader initiative, started by the Broad Institute in Boston and the Sanger Institute in Cambridge, to develop the infrastructure and principles for genomic data sharing on a global basis.

Information databases: To pool knowledge of relevance to all participants, in areas such as toxicology profiles of drug molecules. ChemSpider, a chemical structure database run by the Royal Society of Chemistry, brings together core data on 34 million chemical structures, using data from 494 sources.

Networking platforms: To enable companies to access skills in areas like medicinal chemistry from experts wherever they are in the world.

is turnover of key staff that originated the partnership. With the growing number of partnerships now being forged, there is a real risk of 'consortium fatigue', unless each partnership is managed to deliver against its goals.

There is also scope to open tough problems to all-comers. Crowd-sourcing is becoming increasingly popular in life sciences. Innocentive has become a place on the Web to trade problems and solutions across a range of sectors. Some companies, like Lilly, have had success externally sourcing problem-solving expertise, in their case through their Open Innovation Drug Discovery Program. Similar approaches have been used in identifying optimal synthesis or manufacturing routes for complex molecules. Artificial intelligence will be an additional tool to find and exploit unused IP and other sources of expertise, on a global basis.

The alternative definition of open innovation—eliminating the use of patents—is much more controversial, as patents are currently the principal means of protecting the heavy investment companies make in subsequent product development. The prospect that patents provide of time-limited market exclusivity is a very effective driver for companies making high-risk investments and driving development programmes forward

Box 5.2 Guidelines for successful open innovation collaborations

- Clearly define the opportunities and potential benefits and risks (including any conflicts of interest) for each partner. Be strategic rather than opportunistic.

- Discuss and agree who contributes what to the venture and put in place management arrangements to ensure these are secured.

- Carefully define—and then monitor against—quantifiable objectives and outputs.

- Design a customized IP approach and agree it with all partners upfront. If necessary involve an 'honest broker' organization to hold the consortium's IP, as for example the Foundation for the National Institutes of Health (FNIH) does for the Biomarkers Consortium. IP policy is not as challenging as often supposed, as there are tools available like patent pools and 'protected commons' (IP shared between a group of participants).

- Agree how any other value, including commercial benefits, will be distributed.

- Continuously communicate with transparency throughout.

- Define how, when, and under what circumstances the collaboration will end.

- Have a neutral convenor, if this assists formation or decision-making through the project.

- Have a robust review process to keep the project on track and make go/no-go decisions.

- Remain flexible in budgeting and other arrangements, recognizing that science and its outputs will evolve.

with urgency. However, if a domain needs to be tackled by a group of partners we can use the 'protected commons' route: IP shared among partners. Each can then build their own products on the preclinical, precompetitive findings from the consortium.

For some applications we may need to be more radical. Repurposing existing products for new diseases has been a major theme of the last two decades. With the benefit of more insights on disease mechanisms and findings from data analytics product repurposing is likely to continue to be a major source of innovation. Yet once a patent has expired, there may not be an incentive to do the necessary research to gain approval.

To meet this need, the patent system itself can be modified, or the so-called 'data exclusivity' period (during which regulators will not consider applications for generic products based on the originator's data) could be extended. Alternatively, we may need some form of end-market exclusivity enforced at a national market level.

We need to develop and share enabling technology more efficiently if we are to close the T1 gap. All three types of funders—government research councils, charities, and the commercial sector—need to support collaborative, pre-competitive efforts to crack the problems no one organization can solve. But we need to do so without diluting the tremendous drive that commercial companies bring to subsequent specific product development.

Move decisively to a more adaptive approach to development, trial, and approval design

Most people involved in drug development now recognize that the established model for taking candidate products through to regulatory and reimbursement approval has become inflexible and overly burdensome. In a *Lancet* publication in 2010, I laid out some of the elements that needed to come together in a new flexible blueprint[22]—see Figure 5.1.

This blueprint is intended to tackle a central sustainability problem in life sciences: the high cost, lengthy time, and high attrition rate of clinical development. Without approaches to resolve this, it is impossible to imagine the costs, timescales, and prices reducing to make the enterprise sustainable in the long term.

Fortunately, we now see regulators on both sides of the Atlantic showing openness to a faster, more flexible approach, albeit in slightly different ways. In the USA, the FDA has introduced a series of measures: fast-track, accelerated review, and breakthrough designation routes. We should recall that it was the HIV imperative that began this process; medicines for cystic fibrosis, other rare diseases, and cancer have extended it in recent years. All these options have been taken up enthusiastically, and important new medicines have now emerged from all of them.

In Europe, the last few years has seen moves in the same direction. The EMA introduced an 'Accelerated Assessment' designation that reduces review time for products of particular public interest from 210 days to 150 days. 'Conditional Marketing Authorisation' accepts lower pre-approval evidentiary standards of evidence, subject to gathering further post-approval data. 'Marketing Authorisation under Exceptional Circumstances' can also apply to products for which robust scientific data cannot currently be collected, for example for practical or ethical reasons. Again, there is the expectation of future studies to fill in the evidence gaps.

In addition, there have been exceptions introduced at a national level, typically allowing pre-approval use for compassionate reasons in areas of high unmet need. The French Temporary Authorisations for Use (ATU) and UK Early Access to Medicines Scheme (EAMS) are examples.

When EMA conditional approval was introduced, it tended to be seen and used as a kind of 'consolation prize' for medicines whose evidence base was seen as insufficient for full approval. However, the conditional approval regulation has set the stage for a more significant step on the part of the EMA. The agency is now piloting 'adaptive pathways', in which early cross-stakeholder dialogue (including patients, HTA agencies, and payers) establishes *an evidence generation plan* designed to satisfy all of them[23]. It incorporates the notion of 'progressive authorization', in which conditional approval is initially granted for the patient group most likely to see a positive benefit/risk. The gathering of data from this group can then pave the way for extending the licence subsequently to a broader population.

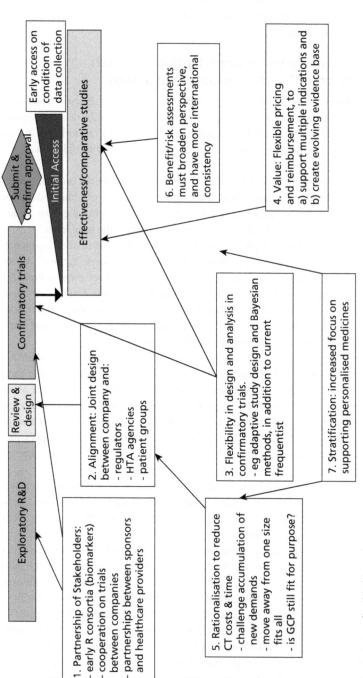

Fig. 5.1 New flexible blueprint.

Early examples are likely to be in diseases that are relatively rare and lack an effective treatment, but the concept is, in principle, extendable to more common diseases where there is a basis for 'stratification' into different patient groups.

This move to 'adaptive pathways' is clearly a major step forward. However, it requires, in my view, an 'adaptive mindset' on the part of all stakeholders, right through to the payers in the health system. This mindset has a number of elements.

First, we need a shift from a regulatory and reimbursement process that assumes we are pitting the efforts of wily and untrustworthy innovators against regulatory and reimbursement 'watchdogs', to one in which there is collaborative design of the evidence package needed to secure approval and reimbursement. Of course, there will be always be some wily innovators, but they will find that the penalties for trying to (or appearing to) hide data or deceive regulators have become prohibitive.

Second, we must move from insistence on a definitive, one-time 'verdict' on a potential new treatment to a recognition that any promising therapy needs to go through progressive testing. This should be on a steadily increasing population of patients, in order to gain a fuller understanding of where it does, and does not, offer a positive benefit/risk ratio. This process better reflects the reality that the professionals know today: that it is impossible to do trials that definitively prove a medicine is 'safe' for all patients, or that the benefit/risk can be precisely known. Only when millions of patients have received the drug do we have a firm fix on this.

This is the basis of 'adaptive licensing': giving preliminary or conditional approval to a medicine on the basis of trials in smaller numbers of patients, with much larger numbers being monitored in real time after this approval is granted. Armed with this additional evidence, regulators can later allow wider populations to receive the drug, or—the converse—use feedback from real-world data to more closely restrict use to those for whom the benefit/risk ratio is highest.

Third, as we will see later, very soon the majority of trials will become trials in 'precision medicine'. Rather than trials seeking the broadest possible approval, products will have been designed on the basis of specific disease mechanisms and markers. Only those patients whose genotypes or phenotypes suggest they will respond will be entered into trials. Trial centres with registers of patients who have been pre-screened for relevant biomarkers will have a major advantage in enrolling patients. Such cohorts of patients will be monitored after formal trials have concluded.

Putting all these three elements together—stakeholder collaboration, progressive review and authorization, and precision medicine—we have the basis for a new process that is being called 'Medicines Adaptive Pathways for Patients' (MAPPs). See Box 5.3 for more detail[24].

Part of the adaptive mindset will be the more active role for the patient perspective. Their views will be sought on what benefits (or risks) matter most to them, and how best to collect outcome data. Discussions between all the stakeholders would take place early in development, ideally in one room, so that each can see the trade-offs inevitably required between cost of development, time to launch, and completeness of data.

There has been some promising experience of this general approach under the auspices of the NEWDIGS (NEW Drug development paradIGms) group at MIT.

Box 5.3 The MAPPs Model

♦ A joint evidence package design step before any significant scale of trials is undertaken, involving all the relevant stakeholders: sponsor, regulator, HTA agency, clinician, payer, and—perhaps most significantly—patient representatives

♦ An initial conditional approval for a well-defined patient population, based on the demonstration of a significantly positive benefit/risk ratio for that population (albeit with a greater degree of uncertainty than delivered by a conventional pivotal trial)

♦ An interim pricing and reimbursement agreement that is then tested in post-approval real world effectiveness studies

♦ Adoption and use under conditions that ensure that the drug reaches only the defined patient group under the supervision of a specialist clinician

♦ Tracking of effectiveness, safety, and value in either the whole population initially treated, or a very substantial sample. This may be done in parallel with a formal confirmatory RCT study

♦ Enlargement (or curtailment) of the treated population once adequate evidence has been collected in the initial group, in one or more subsequent licensing events

♦ Adjustment, if required, to the pricing initially agreed, to reflect the real world experience

Companies have confidentially shared details of a product under development and discussed the development strategy. This was one of the experiences that led directly to the EMA's pilot programme announced in 2014[25].

The task of narrowing the uncertainty associated with a new product, without disadvantaging patients, calls for *creative approaches to trial design*. Especially when diseases are rare and serious, the conduct of normal extended RCTs may anyway be inappropriate. As we saw in the example of metastatic melanoma, exposing a proportion of patients to a regimen known to be ineffective can be regarded as unethical. It may be better to conduct the confirmatory studies—post-initial approval—on an observational basis, for example using historical case controls. Patients need to be at the heart of these discussions also.

Likewise, where conducting comparative studies is costly or we believe we have a basis to assign specific medicines to specific groups, we need to follow the approach pioneered by the iSpy trials on breast cancer and the Matrix trials on lung cancer. Parallel treatment arms, perhaps acting as each other's control. As genomic medicine spreads in its impact, we should see many more approaches like this.

Data analytics can help empower this new model of development, both in the identification of patients likely to respond, selecting them for early trials and also the tracking of patient response after conditional approval is granted.

Once trials are being planned, digital tools can dramatically improve productivity. Trial feasibility can be checked and potential patients can be identified through databases. Websites, linked to patient groups, can help connect to specific patients, whose eligibility can also be checked online. Once the trials are underway, we can also replace human 'clinical research associate' visits and their checking of clinical trial sites and records with electronic surveillance. We can couple these with artificial intelligence tools that can identify suspect data points that need investigation.

As we pointed out earlier, the historical model for development grew out of steadily increasing regulatory requirements. The needs of HTA agencies and payers were grafted on at the end of the process. In most cases, they add another 1–2 years to development times, as they often require additional studies or analysis, and the building of economic models. As we redesign development, we must grasp the opportunity to build economic evaluation, pricing, and reimbursement needs into the new model from the outset. Both HTA agencies and payers need to be fully engaged with MAPPs if it is not to result in a tragic irony—earlier approvals of medicines that are then refused reimbursement. Instead of seeing regulatory and reimbursement processes as sequential, they must operate in parallel. Australia has run them in parallel for the last 4 years.

Many HTA agencies attempt to make their processes as objective and quantified as possible, in part to defend against legal challenge of their conclusions. Some will only accept RCT data. All expect data to be persuasive: the 'benefit of the doubt' is rarely given to the expensive innovation. This has led to an appearance of false precision, in cost per QALY calculations stated to several significant figures, when the real precision is perhaps plus or minus 30%. It has also led to the introduction of a series of exceptions dealing with situations in which the 'objective' measures seem to underestimate the real value of a product in the eyes of clinicians or society.

The more limited data implied by this accelerated model introduces, of course, *greater uncertainty into the HTA process*. This has the potential to worsen this 'quantified value' problem, if we fail to agree a different approach to the value question at the point of initial, conditional approval. One option would be to make this initial assessment simply a 'sense check' of the sponsor's proposals. A more rigorous examination based on trial and real world data will take place at a later stage. Even so, we must deal with a fundamental dilemma at the point of initial approval. If it is for a subpopulation in which the benefit/risk is expected to be highest, then a sponsor might reasonably expect a premium price. However, the HTA agencies and payers will argue, equally reasonably, for an 'uncertainty discount'. Sponsors will be reluctant to agree to such a discount, given the fact that prices in price-controlled markets (most of Europe) rarely if ever increase, and low prices feed into international reference price schemes.

One option would be to create a 'volume contract' for the intermediate period, while the product is being evaluated. 'Coverage with evidence development' is already a well-recognized concept, even if so far rarely used.

In any event, our new, more streamlined development path needs to collect 'value' evidence for reimbursement decision-making, and to do so with international validity. We need to recognize that—while the reimbursement processes we have today have evolved country-by-country—all markets have to make the same basic judgment: how much better is a new treatment than the current standard of care? We need to *converge the 'value' evidence collection across countries*, to simplify and speed that aspect of development. Individual countries and payers will, of course, still want to set or negotiate prices, but this can at least be built on a common evidence base.

Some progress in developing common HTA approaches across countries was made over the last 2–3 years. Six product-specific multi-country workshops were held in the areas of diabetes, breast cancer, melanoma, Alzheimer's disease, and antibiotics, under the auspices of Tapestry Networks[26].

There are, of course, also clinical management challenges to address. The staged entry of a new medicine inherent in MAPPs will require *prescribing controls*. The 'label' initially given by the regulators will specify the sub-group to be treated. However, there will be pressure on doctors, particularly on life-threatening conditions, to prescribe to other patients off-label. That would have the potential to undermine both the acceptability of the MAPPs approach to regulators and the quality of the data collected.

The *ethics* of how patients are involved, both at the point of conditional approval and in subsequent studies, also needs very careful handling. A particular dilemma would occur at the point when patients are recruited for a post-approval study involving a control arm, if there is an alternative of simply taking the medicine under conditional approval. This may require recruitment of subjects for any post-approval studies to have been completed prior to conditional approval. Otherwise patients will appear to have a very unequal choice between being prescribed a medicine thought to be an advance for their disease, or going into a trial with a 50% chance of not receiving it. It would be hard for the ethicists to see this as equipoise—a balanced choice between two alternatives.

Patient engagement will raise a number of other issues. The idea of stratification of patients on the basis of biomarkers may not be automatically understood and accepted. Studies have demonstrated that patients may interpret this as an attempt to exclude them from treatment from which they might benefit. In many ways, it may be better to approach the conditional approval period as if it were a clinical trial. Patient workshops confirm that the greater uncertainty will probably be regarded as akin to that involved in a trial (with the difference, of course, that there is no 'control' arm)[27].

We have already done work with patients and patient organizations on how MAPPs would be viewed. Some of the issues that have surfaced are contained in Box 5.4.

Discussing the issues associated with the adaptive model directly with patients also reveals that—from the MAPPs perspective—not all diseases are equal. Not surprisingly, patients with slowly deteriorating conditions with some existing therapy are less likely to risk additional uncertainty than those facing a fast decline and death. While patients vary widely in their understanding of the existing system, all are uniformly frustrated with its slowness and open to radical changes. However, they want to be treated as equal partners in change. They want to trust their doctors, but expect transparency and fairness in return.

We need also to examine the *legal implications* of the new route for both sponsors and prescribers. Legislation in some countries already gives doctors discretion to prescribe unlicensed medicines on a 'named patient' basis. However, more widespread use of a medicine carrying greater levels of uncertainty—albeit on an 'conditionally approved'

Box 5.4 Patient reactions to the MAPPs concept

♦ The need to communicate respect and reflect this in open communication and the sharing of information throughout the process

♦ The recognition that patients (and their carers) vary widely in their knowledge of the current process and in awareness of the considerations involved in MAPPs

♦ The need for trust in both the process and the professionals involved

♦ A strong focus on the value of empowering the patient to be part of the decision making process right from its conception, moving away from the 'passive patient' model to 'patient leadership'

♦ An understanding that, for some patients, their motivation is gaining knowledge for future generations, rather than their own immediate benefit

♦ The need to explain, in as much detail as possible, the potential for benefit and for risk, and the level of certainty/uncertainty

♦ The importance of a proper consenting process, reflecting the differences in the situations of patients with different diseases

♦ The need to be flexible in study design, and to avoid a placebo arm when this offers no potential for benefit

♦ The need for maximum fairness and transparency in determining patient eligibility for inclusion

♦ The importance of putting the process in the hands of specialist doctors and having them act as a single point of contact

♦ The need for patients to consent fully to data sharing (although personally identified data should only be available to the professionals caring for them)

basis—could result in side-effects (or worse). In principle, these could form the basis for a legal challenge, especially in litigious environments. Ultimately, I believe we need reform of the legal framework that opens up companies to being sued for problems (e.g. side-effects) that no-one knew existed. Otherwise, company legal departments could torpedo the new model. A state- or industry-funded compensation fund may also be a way forward.

Finally, we need to work on *public and media understanding* of the adaptive approach. It can easily be seen as an attempt by sponsors, abetted by regulators, to expose patients to 'untested medicines'. The media never tire of 'human guinea-pig' headlines. Avoiding this will require careful explanation of some challenging concepts. One is the difference between increased risk and greater levels of uncertainty. Both risk and uncertainty are concepts that are difficult to communicate in popular terms. The public—and the politicians who respond to them—need to understand that MAPPs involves not greater risk but increased uncertainty about *both* benefit and risk. This will require design of new communications tools and media briefings, in which patients that could benefit should play a prominent role. The adaptive approach must be seen in the context of its potential to meet urgent patient need.

Currently Europe (or some countries within it) is in the lead in shaping the adaptive pathway. At the time of writing at least ten medicines are being piloted in the 'MAPPs' process. The philosophy on the part of the EMA (that is leading the pilots) is of 'learning by doing'. They rightly realize that trying to anticipate and manage all the possible consequences of a new route would delay its introduction by years.

A refined version of this EMA process, adapted for the UK situation, is shown in Box 5.5.

Of course, there will be examples of products for which the early promise is unfulfilled in wider use, either because of poorer efficacy or increased side-effects. Protocols for withdrawing or more tightly restricting use will be needed.

Most of these issues are minimized if patients are closely defined, treatment centres and clinicians are likewise tightly controlled, and data requirements fully met. These operational issues will be critical to the success of the new pathways. The price of greater flexibility in the process will be tighter control of its implementation.

Box 5.5 Accelerated UK access pathway

- Identify appropriate innovations for accelerated access: high unmet medical need, strong initial efficacy data, 'transformative' impact on patients and/or health system efficiency.

- Hold cross-stakeholder workshop (involving sponsor, clinician, patient organization, regulator, HTA agency, and payer) to develop evidence generation plan, incorporating agreed benefit/risk measures and threshold.

- Develop a priority clinical trial plan around required evidence and defined patient population, ensuring an adaptive trial design is used if appropriate; investigators are identified and adequately incentivized; patient recruitment utilizes available registries and/or patient organization membership; biomarkers for inclusion are available.

- Conduct trials and submit dossier for conditional approval for defined population.

- If benefit/risk threshold exceeded: regulator to issue conditional approval and identify additional data collection required for full approval; HTA/agency and payer to approve a reimbursement arrangement via a 'commissioning with evaluation' route.

- Identify centres for conditional use and the data collection regime required to confirm or refine B/R assessment.

- Following the defined conditional use period, submit additional data package to regulator, HTA agency, and payer for expanded approval and normal reimbursement arrangements.

The adaptive approach may also be applicable to high-cost medical devices. Often it is difficult to complete full-scale randomized trials where long-term survival is the end point. However, significantly positive results in a small number of patients could enable 'conditional licensing'. As Carbonnell et al. have argued, for the case of Canada, 'Conditional licensing may be granted to new moderate- or high-risk medical devices … when there is reasonable assurance that the device is safe'[28].

Then, as already is the case in France, the assessment authority can plan to re-review the data a few years later. At that point, they can decide to continue, restrict, expand, or withdraw the product from the market, on the basis of performance. Of course, should the results be disappointing, it is unlikely to lead to surgical removal of the device from the initial patients!

One of the most important aspects of the assessment of a new technology, be it drug, device, or diagnostic, is the impact on the overall pathway of care. It is important to assess the health and economic impact of the total change the technology drives, rather than simply compare it to the specific technology it replaces (as we saw in the case of RA).

This move to adaptive development and licensing—for drugs or devices—is not a step in the dark. Australia introduced its Managed Entry Scheme in 2011, with interim pricing arrangements during the period of 'coverage with evidence development'.

Longer term we need to create ever-greater *harmonization and mutual recognition* across different jurisdictions. The 'International Council for Harmonisation of Technical Requirements for Pharmaceuticals for Human Use' (ICH) process has focused on common standards, which is helpful, but multiple reviews to establish the same facts are a waste of money and time. Some useful steps are underway, like the Medical Device Single Audit Programme (MDSAP). This is a pilot to conduct a single manufacturer audit across Australia, Japan, Canada, and Brazil.

It will already be apparent that new, faster models of development are only possible if we have the *data infrastructure to collect post-initial approval experience* quickly and reliably. In the case of rare diseases, it may be necessary to do this across several countries. So, just as we should be more adaptive in the overall development process, we need to explore new ways to collect evidence of efficacy, safety, and value.

Adaptive approaches to clinical trials

The *clinical trials* themselves can be more flexibly or adaptively designed, and real world data can be incorporated in regulatory and reimbursement decision-making. Patient registries will need to be a growing feature of the scene.

Double-blinded RCTs have been the 'gold standard' of clinical evidence since the early 1960s (see Box 5.6).

RCTs are clearly the most rigorous and objective approach to clinical evidence. However, as someone once pointed out, it is like the result of a standard 'urban cycle' test of a car's gasoline consumption: completely objective but largely unrepresentative of real life. As Mike Rawlins, chairman of the UK regulatory authority (the MHRA) has pointed out, RCTs are just one of a wide variety of clinical study designs[29].

In RCTs, there is no concept of accumulated learning. Under the 'null hypothesis' each successive trial needs to prove conclusively that the product offers clear benefit/

Box 5.6 Randomized controlled clinical trials

- ◆ A well-defined population, meeting rigorous inclusion and exclusion criteria, is randomized systematically so that neither patient nor doctor knows whether they are on the treatment arm or the control arm of the trial.

- ◆ A well-defined protocol to ensure treatment—and the conditions under which it is given—are as consistent as possible.

- ◆ Defined data are to be collected on a timely basis.

- ◆ A clinical trial supervisory group to ensure that if any intervention in the course of the trial is necessary (e.g. if a treatment is clearly dangerous, or is so much better than the control that it is unethical to continue) then it takes place.

- ◆ Careful, balanced statistical analysis of the results, based on the 'null hypothesis', assuming the trial runs its course.

---

risk to patients. No account of the results of previous trials can be incorporated in the assessment: it's rather like the legal rule in most countries that previous convictions of a defendant cannot be included in the evidence presented to juries, lest they be biased in their decision.

However, accumulating evidence by combining two sets of scientifically derived results does not make the combination less scientific. The barrier to this 'Bayesian' approach, according to senior regulators, is less a question of logic and more that of skills: 'Bayesian' statistics are not understood by all statisticians!

In principle, there is enormous inefficiency in every company conducting separate trials on their individual products for the same disease. Often this duplicative process quickly exhausts the pool of available patients. In cancer, two new types of structured trials are gaining popularity—'umbrella' and 'basket' trials. Both take advantage of the new insights into cancer mechanisms provided by genomics and other biomarkers, as commented above under 'Advance the molecular definition of disease and the application of systems biology'.

In 'umbrella' trials, patients meeting pre-defined criteria (e.g. metastatic melanoma with a BRAF mutation) are randomized between a number of different trials arms, each of which has a different new agent, often from different companies. The goal is to identify the most effective new drug, while keeping the costs of the trials down for all participants.

The US-wide iSpy trials in breast cancer have been widely supported. So far there have been two waves of such trials at the Phase 2 stage, involving several different agents. Patient outcomes are used dynamically to influence allocation of future patients to trial arms[30].

The current 'Matrix' trial in the UK in lung cancer using this approach has been set up by AstraZeneca and Pfizer, in partnership with Cancer Research UK[31].

In 'basket' trials, tumours are no longer defined in terms of organ of origin but the specific molecular profile. So a trial may contain patients with 'breast', 'colon', or 'prostate' cancer, but in each case with (say) a BRAF mutation. The authors of a recent review of how adaptive trials could contribute to a single area of cancer (thoracic) contrast the classic approach of 'trials designed to conclude' with 'trials designed to learn'[32]. This sums up a fundamental change of mindset.

As we consider such changes in trial design, we should also review trial endpoints. In cancer, overall survival has long been the gold standard required by regulators and reimbursement agencies. However, by definition the readout from trials becomes very long, raising the need for predictive indicators. Tumour shrinkage is not very reliable: often treatment can kill some of the clones but allow more dangerous ones to flourish and spread. Progression-free survival is more frequently used, but also can mislead.

Imaging gives us faster readout, not only on the primary tumour but on metastases, and such tools can be used to more quickly identify which patients are beginning to respond.

I have given a great deal of space to the third step of adaptive development. This is for two reasons. In terms of contribution to the overall goal of more treatments, faster and more affordably to patients, this series of changes has probably the greatest impact, and it is under very active study. It is also a particular personal interest.

At this point, we should check that we are addressing the many T2 issues we described in Chapter 3. From Figure 5.2 we can see that adaptive development, including novel trial design, can address the first set of issues. Cross-stakeholder dialogue on evidence requirements, in a 'safe harbour' environment, can deal with the second. A greater level of collaboration within trials and in building infrastructure will focus competition on where it belongs—the quality of the innovation. However, the fourth area, the need for a more joined-up approach across agencies, both at a national and international level, will not happen automatically. We need to create policy vehicles for this. We will return to this subject in Chapter 6.

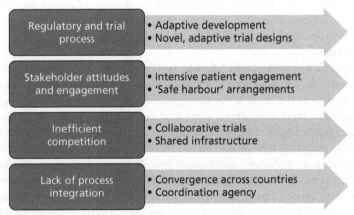

Fig. 5.2 'T2' issues and solutions.

Create new reward and financing vehicles for leading edge innovation

Financing innovation is becoming a tougher and tougher problem for health systems. They are themselves caught in a sustainability problem, as I pointed out in an earlier book[33]. The underlying problem is an inability to fund care as it is delivered today in the face of trends in demographics, disease incidence, and patient expectations. These forces require a radical change in the systems themselves. For unreformed health systems, continuing with today's inefficient pathways of care, any externally supplied innovations—whether drugs, devices, or diagnostics—are the most obvious targets for cost control. So most systems have, or are building, mechanisms for volume or price control, or both. These mechanisms are often applied even to those innovations that save money for the system in the long term, because the products cost more in the short term.

Therefore increasingly we see HTA agencies determining that products that may have been developed over a decade—often at development costs of hundreds of millions of dollars—are not cost-effective and will not reimbursed. But we also find HTA-approved products (like HCV treatments) being refused by payers, who say that they are simply unaffordable within their current budgets. This situation will only get tougher, as expensive targeted medicines come forward. They will be particularly acute for advanced gene, cell, or tissue therapies that could be one-time cures, but come at a very high price.

It is in everyone's interest to explore creative alternatives to the current logjam—one that is most acute in Europe but is also a potential challenge elsewhere. But we should start by analysing the source of the tension.

The current business model for pharmaceuticals, and indeed most life sciences products, is simple: establish patent coverage, secure marketing approval, set a price per unit (often in negotiation with the payer), and then maximize revenues by vigorously promoting the product to prescribing professionals while the monopoly lasts. From the industry standpoint, this plays to their strengths: global development, marketing, sales, and distribution. It is administratively simple for the doctor, who simply prescribes and for the payer, whose bill is just the number of units times the price agreed. The intervention of the payer into this process is often described as a denial of 'market access', a phrase which implies the doctor is the customer, and the health system is not.

This formula has worked well for many years and in most circumstances, but can have unfortunate consequences. Companies are tempted to maximize revenues by promoting outside the 'label' set by the regulator. When a new indication is granted by the regulator, there is usually no scope to change the price, even if the value in the new indication is higher. Payers are understandably cautious when negotiating the initial price, since they have little or no control over how much of the product is used. And no-one in the industry–doctor–payer triangle is accountable for ensuring that the product prescribed is actually used and the patient receives the benefit intended.

As we think about adaptive development, it is time to also consider adaptive business models. The traditional business model may be perfectly adequate for low-cost products to mass markets, but it does not work well for specialist products that need to be well targeted but then often move from niche to niche (as for many cancer drugs); for products whose use needs to be severely constrained (as we saw in anti-microbials); and especially for products that are extremely expensive yet cost-effective in the long run (as in HCV anti-virals).

There have been experiments in so-called 'risk-sharing', linking the reward to the company to the outcomes actually achieved. However, in the UK for example, such schemes were quickly seen as an administrative nightmare: outcome data was not being routinely collected. The schemes became 'patient access schemes' and quickly reduced to demands for a confidential discount (eliminating risk on both sides). Thereafter the model operated as before.

There are a number of ways in which the reward system could be adapted, to better align incentives and create greater sustainability in circumstances in which the traditional model fails.

Firstly, to agree a formula that decouples reward from volume driven by promotion. This is almost certain to be necessary in the case of new generations of antibiotics that need to be kept in reserve for resistant cases. Fixed rewards, such as advance purchase commitments or 'prizes' for successful product development, are two options. This mechanism was explored in a recent Chatham House document on 'learning from other industries'[34]. The formula presupposes that the successful initial innovators have a fully effective product, or that the reward can be shared with those with better products that come to market later.

Secondly, to link all or part of the reward to the innovator to the improved health achieved. This would have the advantage of incentivizing the innovator to ensure its product is properly used. This means investing in the infrastructure to collect outcome data, which is becoming easier as records become predominantly electronic and their analysis more commonplace.

Outcome-linked rewards would be useful, for example, in those situations (like advanced cancers) in which products are used in situations beyond those in formal trials. Where ethics preclude initially testing the drug on early stage patients, clinical trial results are usually on refractory patients, for whom the drug often extends life by just a few weeks. As the usage spreads to earlier points in the course of the disease, better impact is often achieved, but often the price calculation stays linked to the original clinical trial results. Herceptin is a good example. If the longer survival achieved from earlier use were reflected in rewards, this could overcome this problem (while of course setting some challenging data collection tasks).

The Swiss company Novartis is planning to test an outcomes-based pricing model on a new drug Entresto, approved in 2015, for congestive heart failure. Payers will be given an option to pay a discounted price but then reward the company additionally on the basis of reduction in costly hospital visits. The rationale for the company is that market penetration under normal pricing arrangements will be difficult, as the drugs that the new product replaces are relatively cheap. We will see more formulae of this kind as expensive new products meet reimbursement resistance.

A *third* formula may be appropriate when a technology shows very high promise on the basis of limited human trials, and conditional approval for a sharply defined group of patients is sought and given. Here there is a period of proving out the potential shown in the initial trials. As we saw earlier, the innovator may expect a high price, because of the high impact and small pool of patients; the payer may expect a low price, because of the limited data available at the time of initial approval.

While the product is being further evaluated, a 'reimbursement in research' formula may be most appropriate. This could, for example, be in terms of a fixed research contract calculated on the basis of the numbers of patients expected to be treated during this period. At the point at which full approval is given, the HTA system will have more extensive data to evaluate.

However, none of these three formulae tackle the dilemma of products in the 'high upfront cost—very long payback' category. The HCV anti-virals and future gene and cell therapies are examples, as are costly implantable devices. Long-term calculations may show the intervention is very cost-effective, but the short-term cost to the health system is deemed unaffordable.

Of course, as some in industry would point out, affordability is a matter of relative priorities, especially in government-financed healthcare. These kinds of innovations should simply be placed alongside other funding decisions that governments make—high-speed train networks, defence capabilities, etc—that are regarded as long-term investments for the good of their populations. However, that argument ignores the reality of the way in which healthcare is funded, both publicly and privately. In the public sector, budgets are given to commissioners or providers year by year. The fact that they will save money in the long term can seem academic to hard-pressed health officials with limited annual budgets. And in private healthcare, when patients switch from one insurer to another, the investment made by the first is actually recouped by the second.

Triggered by the Sovaldi experience, thought is now being given to creating *novel financial instruments* that cover the cost (and therefore bear the risk) of the initial investment and are paid back from the long-term returns achieved. There may be a role for bonds issued by banks, or insurance intermediaries in these situations. Such mechanisms have been used in other parts of healthcare and infrastructure development, where major upfront investments are made for a long-term benefit. However, the design of these will be a complex business, as they need to take into account judgments on long-term product effectiveness, the impact of non-adherence on this, the entry of competitive products, and several other moving parts.

Finally, in all the evaluations we make, both interim and 'final', we need a *broader definition of value* than applied in many countries. Value, and therefore pricing, needs to reflect not only the direct health impact, in terms of benefits and costs, but also the societal impact of extending or restoring the ability to work, or releasing carers from the need to provide constant care to patients who are relatives. We also need to find a way to reflect the level of innovativeness that a new product represents, especially when it opens up a whole new avenue of treatment in a severe and costly disease (as we saw in RA). This is challenging to implement, but as a society we need to reward boldness in pursuing new approaches and mechanisms.

My overall point is this. If we are to ensure that reimbursement is not a barrier to valuable innovations reaching the patient, we need to be more imaginative and adaptive in designing the business relationship between innovator and payer. And in the process, perhaps, provide incentives to the system to make sure the benefits of the innovation are actually delivered. Such schemes—like adaptive development pathways—will probably initially be offered as options to the innovator or proposed by the innovators themselves.

Engineer tools and systems for faster and better innovation adoption and adherence

Despite all the science that goes into the discovery and development of medical innovations, the science of their adoption by doctors and their use by patients is remarkably primitive. Adoption is largely based on energetic promotion to physicians by commercial companies. Once prescribed, the use or adherence by patients is largely hit-or-miss.

With a combination of behavioural science insights and digital tools, we can do much better. We must understand the needs and concerns of both clinicians and patients and tailor our strategy to these. We can also use the resources of both public and private sector organizations interested in the appropriate use of innovations, once their clinical value and cost-effectiveness have been established. There will still be a place for classic promotional and educational mechanisms but we can also *engineer* for uptake and adherence with the tools now available.

An innovation will often trigger a change in the pathway of care, or the algorithm that the doctor applies. We need to apply engineering thinking to the change. How can we 'wire in' the new product and the new pathway while still allowing the doctor some discretion?

Adoption and diffusion of innovation among doctors

Most doctors regard themselves as independent professionals, able to make their own judgments about which treatment to prescribe. If we are to engineer faster, broader adoption of proven therapies, we must understand how they wish to learn about innovations and invest in implementation science based on a deeper understanding of behaviour and its motivation, but then make it easy to change.

There is a huge management literature on change in organizations, much of which is irrelevant here, as it applies to situations where leaders decide and employees implement. The key difference in healthcare is that we are dealing with highly trained individuals many of whom make a wide range of treatment decisions: we are interested in changing these one by one, as innovations and evidence about them emerge. And, as we have seen earlier, the volume of medical literature, even over a narrow specialism is now so large no one doctor can keep current with it.

We have discovered that doctors are open to a range of channels through which to learn about innovations[35] (Figure 5.3). However, learning from their peers is by far the most important source. The medical representatives sent by companies are (or certainly are perceived to be) much less important.

However, innovators typically rely on medical representatives to inform individual doctors and address their questions. This can have some unfortunate consequences: it is sometimes sheer marketing muscle or high pressure salesmanship that determines the treatments that flourish, rather than their intrinsic merit. And the health system

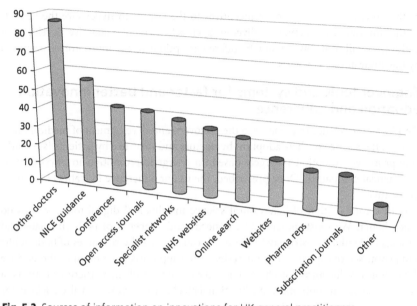

Fig. 5.3 Sources of information on innovations for UK general practitioners.
Reproduced with permission from Nesta. *Which Doctors Take Up Promising Ideas?: New insights from open data*, http://www.nesta.org.uk/sites/default/files/which_doctors_take_up_promising.pdf, accessed 01 Oct. 2015. Copyright © 2014 Nesta.

sees its role as controlling and challenging this adoption process, rather than assisting or even driving it.

What are the alternatives? Certainly evidence-based *guidelines* are among them, such as the clinical guidelines developed by bodies such as NCCN (National Comprehensive Cancer Centers Network) for cancer treatment in the USA, or NICE in the UK—as long as such bodies can respond rapidly enough to the flow of innovations. But too often these guidelines are in lengthy reports that gather dust on the surgery shelves.

We need to design modern *decision support systems* that fit comfortably within clinical practice. As an imperfect analogy, pilots have a 'heads-up' display of key operational factors so that they can make their decisions rapidly without taking their eyes off the sky or runway in front of them. We need to engineer something similar.

Access to the latest information on a new treatment needs to be at the doctor's fingertips—or more precisely on the screen in front of her—as she makes her decisions. She needs to know where the treatment sits within the pathway of care, along with diagnostics necessary to make the right choices. Therapy options, their implications, benefits and risks, and any patient-specific factors need to be spelled out in a way that is customized for the doctor and the situation. This information needs to include the outcomes of previous decisions by other doctors in comparable cases.

The speed of adoption of innovation in the UK's NHS is one of slowest across major countries. NICE assessment, followed by a series of evaluations at different levels of the NHS, conspire to defeat even some of the most professional marketing organizations.

There are exceptions: specific patient management steps, screening and diagnostic tools that are included in the General Practitioners' contract (via the so-called QOF—Quality Outcomes Framework—metrics) are remarkably well adopted (more than 95% across the NHS). Achievement of each 'QOF' goal is linked to payment. *Financial incentives* can work wonders.

To address the problem more broadly, the NHS set up 'Academic Health Science Networks' (AHSNs) to take a rigorous approach to spreading evidence-based innovation. Each of the 15 AHSNs has an area of the country of around 2–4 million people, which makes the task more tractable than a set of top-down dictates. The experiment is nearly 3 years old and early results are promising, with, for example, public–private consortia to promote the optimal use of products. The South London network has pioneered one of these in the diabetes area: the 'Diabetes Improvement Collaborative' aims to increase the adoption of insulin pumps, from the current 5% to the 15% recommended by NICE[36]. Such partnerships require careful matchmaking, as there are potentially conflicting interests, and certainly cultural gaps to bridge.

Where commercial interests could conflict, as in the area of product choice, partnerships can still be formed between clinicians, commissioners, and pharmacists. An example is a programme 'Right Insulin, Right Patient, Right Time, Right Dose', optimizing the use of different types of insulin at different stages of diabetes [37].

Where the changes are complex, such as the adoption of a new type of medical device through a surgical procedure or a digital tool for managing a medical practice, many more factors enter into the equation—as we saw in Chapter 3. Does this innovation disrupt the normal flow of the doctor's practice, require him to learn a new set of skills or remember a new and complex set of facts to apply it properly? Is the necessary technical or administrative support available? Are there economic consequences for the doctor—positive or negative—resulting from a switch to this innovation?

These more complex challenges represent the opportunity to practise 'implementation science', a relatively new discipline defined by the NIH as 'the study of methods to promote the integration of research findings and evidence into healthcare policy and practice.' The incorporation of evidence-based medicine into routine practice and its spread through the system is a major focus.

To achieve implementation, there needs to be opportunity, capability, and motivation (rather reminiscent of 'opportunity, means, and motive' in criminal detection). If we want to intervene to speed the process, we have a wide range of tools to use, from the softer tools of education and training through to coercion, which is a last resort with medical professionals, in any event (Figure 5.4)!

We need now to go beyond 'implementation science'—the study of how things are, or might be, implemented in health systems. We need 'implementation engineering': constructing the system to *ensure* implementation, once an innovation is demonstrated to be value-adding, but in a manner that respects the medical professional's experience and the individuality of every doctor–patient encounter.

Building on the peer-to-peer principle, there is a major role for innovation champions, who can spread value-added innovations to colleagues. Mechanisms such as local area seminars and incentives to reward the champion role would assist that process.

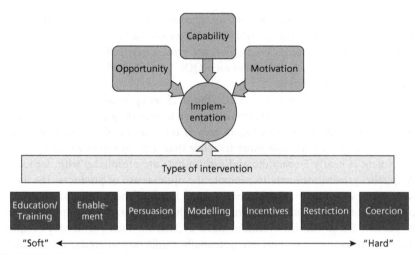

Fig. 5.4 Implementation science.
Adapted with permission from Michie S, van Stralen MM, and West R. The behaviour change wheel: A new method for characterising and designing behaviour change interventions. *Implementation Science,* Volume 6, pp. 42, Copyright © 2011 Michie S, van Stralen MM, and West R.

However, the most powerful means of driving uptake of innovations—and the new pathways of care that they often enable—is to explicitly commission them: link payment for care to the adoption of the new pathway. Once care commissioners are convinced, for example by an analysis by a body such as NICE, that an innovation and its pathway delivers a better ratio of outcomes to cost, it should become automatic to commission care in those terms. Of course, the total cost has to still to be managed, which requires better horizon-scanning data and economic dialogues than we have today.

It is clear there is no silver bullet: training, decision support, incentives, or commissioning could all have a role in spreading an innovation. But we need to take an engineering approach to the challenge, using a combination of pathway-based decision support, peer-to-peer seminars, incentives, and commissioning. Each speciality will probably require a different engineering approach and mix of tools.

Patient adherence

Turning to the patient, the fact that large numbers of patients fail to take their treatment as prescribed—or even to have the prescription filled—has often bewildered medical professionals. It has frequently been put down to forgetfulness, as if the condition that drove them to the doctor in the first place was suddenly something they could quickly overlook. So the best that health systems and companies could do was to put in place text reminders, calls from nurses, etc.

Recall (from Chapter 3) the two major groups of factors that drive the decision of patients to adhere (or not) to the treatment that the doctor prescribes. The first is the sense of *necessity*. Is this condition something the patient genuinely feels to be serious, in the short or long term, for their health or the length of their life? Diseases naturally fall along a spectrum of immediacy and impact for the patient. Immediate relief of sharp pain is at one end of the spectrum; prophylactic treatment for a symptomless condition such as high blood pressure is at the other. Experience suggests that, in the absence of other prompts, people tend to discontinue health-promoting behaviour with no immediate feedback after around 3 months. Where a sense of necessity is not there, a simple prescription is not good enough. Information on consequences—in whatever form really hits home, for example in a video—is critical.

The second factor, underlying *concerns* that the patient may have about the treatment itself, needs to be analysed and dealt with equally effectively. For medicines or vaccines, this concern understandably often centres on taking a foreign substance into the body. Rumours about mercury in vaccines, or worries about the build-up of chemicals in the body, persuades the patient that the cure could be worse than the disease.

So, as a result of a conversation with the doctor and the use of other aids, we need to implant a 'commonsense' story in the patient's mind: this needs to marry a sense of necessity to act—based on understanding of the disease, its consequences, and the value of the treatment—with a belief that any concerns about the treatment have been understood and answered.

Companies like 'Spoonful of Sugar' and Atlantis Healthcare have now been created to advise pharmaceutical companies and healthcare systems on how to support adherence to products. They find—through patient questionnaires—that patient beliefs and their sense of need and of concern is a very individual matter. Beliefs include disease causes, symptoms, consequences, and likely duration. The patient's body image can also be relevant[38,39].

So the intervention—the new information and education given to the patient—has to be highly customized. The *content* must respond to the individual's needs and concerns; the *context*—in terms of personal circumstances and clinical support—taken account of; and the *channel* for delivery of the message must be appropriate.

The failure to engage in this customized way with the patient accounts for much of the disappointment to date with digital means of managing health and disease.

We are also learning the power of 'gamification'—making adherence to a health-promoting regimen a matter of competition. This can be against the patient's own goals or compared with a cohort of similar people. This technique seems to be effective in maintaining interest and engagement beyond the 3-month 'boredom' point.

So we are learning that not only can and should medicine become more precise, but its delivery must take account of the motivations and issues of individual clinicians and patients. This is a task that must marry engineering with psychology. Engineering is needed to design reliable ways of delivering information, wiring in implementation, and tracking response. Human psychology will tell us how each individual needs to interact with the decisions that face them. In our age of 'mass customization' in other walks of life, we should be equal to that task. If we can design the right content, context, and channels, we can perhaps shorten the average time of adoption by several years and double the probability of adherence.

Develop an infrastructure for real-world data-driven learning

Our historical paradigm for learning about an innovation has been (1) to conduct clinical trials on restricted and unrepresentative groups of patients and then (2) ignore any information from the post-approval use of the product, often on millions of people. For too long, the medical system has been essentially 'fire and forget'. Doctors learn the outcomes of their treatment decisions only if patients re-present, as a result of dissatisfaction. And, of course, many patients simply shrug and suffer when a treatment fails to deliver.

In fact, the outcome from every therapy decision on every patient is a chance to learn more about how it performs in the real world.

The advent of the electronic medical record is now slowly changing this, and there has been keen interest over recent years in so-called 'Real World Data' (RWD). Patients can be studied in much larger numbers. Information is much more representative of the typical recipients of treatment (for example, elderly with co-morbidities, who would never be in clinical trials). The context and conditions are more typical, for example in a primary care practitioner's office, rather than a specialist centre. It can include data contributed by the patients themselves, under the broad heading of 'Real World Evidence' (RWE). (NB: the terms RWD and RWE are often used interchangeably.)

It is still, of course, important to assess an innovation against an alternative—typically the prior standard of practice. Databases will, of course, contain the results of both, but coming to robust conclusions is a challenge. Statistical purists find many sources of possible inaccuracies in judgments made on the basis of RWE. Even if there is an attempt at randomizing patients on the part of the doctor, there will be inbuilt biases in who receives which treatment. Other factors that interfere with the treatment regimen can confound the conclusions. Patients may hardly be impartial contributors of data, if they know what treatment they are receiving. The list of objections is a long one.

Yet there is a similarly long list of objections to the results of an RCT. Patient populations are selected—using many inclusion and exclusion criteria, for their simplicity and straightforwardness, rather than their similarity to most of the patients that will actually receive the treatment post-approval. Such populations are often white males. Patients in both treatment and control arms receive more attention and supervision than is normally the case. And so on.

The right perspective is that both types of data—RCT and RWE—can be hugely informative, but in different ways. The mix will depend on factors such as the ability to find enough patients for a properly constructed RCT and the extent to which such patients are unrepresentative of most of the patients who are intended to receive the therapy.

The early discussion of RWD, Big Data, and so on has focused on our new-found ability to store, handle, and analyse huge databases. Although 'size matters', to ensure a sufficient statistical base, in practice most health insights will relate to specific patient profiles and the use of innovations under well-defined, comparable circumstances.

With the new data handling capabilities, there is an increasing drive to create open knowledge ecosystems, so that experience and outcomes can be pooled across practices, systems, and countries. There are barriers to this, of course—confidentiality and commercial interests, in particular. However, de-identified data pooled in a collaborative fashion in a secure system has the potential for everyone to learn what no one doctor, hospital, or system can deduce from their own limited data set.

So, with the advent of electronic medical records, we have the opportunity to gather and analyse decisions and outcomes on a scale never before possible. In many cases, we still need to overcome technical problems of incompatibility in format, or interoperability between record systems that each contain part of a patient's care. This barrier between primary and secondary care is particularly challenging in many systems, especially where they are under different ownership. But, as we overcome these problems we can turn medicine into an evidence-based 'learning machine'.

What are some of the lessons we will learn from RWE? Firstly, that *some conventional treatments simply don't work*. As the Cochrane Collaboration has found, serious examination of the evidence for a number of broadly used therapies has exposed their ineffectiveness[40]. More than 20 000 medical experts are assembled in review groups to sift through the body of evidence on specific treatments. For example, although there might be an a priori reason to expect that statins could prevent dementia by reducing cholesterol and so improving vascular health, in fact they fail to do so[41].

A second type of learning will be *phenotypic differences that affect treatment outcomes*. It was only after disappointing overall clinical trials for the non-small cell lung cancer (NSCLC) drug Iressa (gefitinib) that it was found that non-smoking Asian women responded much better than the average patient[42]. Armed with 'data discoveries' like this, we can then go back to the laboratory to discover the molecular basis of such differences.

Thirdly, and perhaps most promisingly, we will find *unexpected efficacy signals*—via correlations between treatments prescribed for condition A and the incidence or progress of disease B. We must watch out for confounding factors of course—common underlying characteristics that explain the correlations. And we will want to understand the mechanism. But they provide the basis for prospective trials that can eliminate these confounding factors. An example is the real world data-based discovery that metformin prescribed for diabetes may help prevent cancer[43]. There is a putative mechanism (interaction with growth factor IGF-1), but this finding is still controversial, and there needs to be careful analysis of the time periods for treatment.

However, such data discoveries will prove to be good starting points for both mechanistic research and prospective clinical trials.

Real world evidence comes in several different flavours and can be used for a number of purposes. So-called pragmatic trials test a new product or procedure in a more realistic setting and/or using adaptive trial designs. This approach is being investigated by IMI through a project called GetReal, which is looking at RWE use in both regulatory approval and HTA analysis[44].

In the USA, there is a particular opportunity to access the huge body of health utilization and claims data that has been collected for decades. Large businesses have been built on supplying analysis to both health plans and pharmaceutical companies; however, so far this data has been poorly integrated into the clinical data domain.

As RWE develops, there will be insights on three levels:

- Total population. Like the most important epidemiological finding of the last century, that smoking causes lung cancer, we will find correlations that guide the management of whole populations.
- Stratified groups. Groups of patients that show better or worse response will help target therapies and design future trials.
- Individual findings. Specific 'outlier' cases that emerge from the analysis may provide indications that diagnoses are wrong, or that particular therapies can have unexpected effects in individual cases.

We are amassing both data and tools on a scale never seen before. When coupled to new machine learning approaches, the potential is tremendous. IBM's Watson is already partnering with the US pharmacy chain CVS (with its 7800 outlets and millions of retail and pharmacy customers) to identify people with declining health likely to benefit from proactive, customized patient management. CVS is already talking about integrating this with activity measures from wearable devices[45].

The technology will come into place step by step. However, if we are to realize its full potential we must address some major ethical and public engagement challenges. We will return to these in the Chapter 6.

Engage patients wholeheartedly, throughout the innovation process

We need to leave behind a paternalistic innovation process that treats patients as passive clinical trial subjects and recipients of end products. The future process must be one in which patients are fully involved in therapy specification, development design, approval mechanisms—and provide real world feedback on effectiveness.

Patient input adds value and can sometimes change priorities at every stage in the innovation process. CASMI brought together some UK patient organizations, in conjunction with Genetic Alliance UK, to debate and document how *they* would organize the R&D process. The result was a 'manifesto' with several important differences from the current process—one that was largely developed without patient input[46]. (See Box 5.7.)

Here are some of the most striking differences between traditional R&D perspectives and those of the patients:

Research priorities

There is a very strong orientation among researchers to treatment and cure of the primary cause of the disease; patients often want alleviation of some of the most distressing side-effects either of the disease or current treatments. We should start with the definition of the problem patients face, and this often revolves around the ability to live a normal life while waiting for a future cure.

Benefits

There is an understandable focus in research on benefits that are accurately measureable in clinical trials, but these do not always reflect how patients view effectiveness in each stage of the disease. For example, in some progressive diseases that result in

Box 5.7 Summary of the patient-focused R&D manifesto

Basic research on the causes of diseases

Failure to make progress in many areas results from a lack of understanding of the causes of disease. Investments in disease research need to be more strategic with less duplication and a greater sharing of assets and data.

Benefits sought in new medicines

Research is often 'push'-focused, targeted on what professionals in industry believe are easily measureable performance criteria; patients want a more 'pull' orientation, prioritizing other, 'quality of life' and patient-oriented measures.

Non-drug innovations

Managing patients' lives in the later stages of disease requires technologies that need to be more actively developed, tested, and reimbursed.

Clinical research

There is a major opportunity for creating disease registers (with the support of patient organizations) and linking these to tissue banks.

Regulatory benefit/risk assessments

These should take patients' views much more systematically into account and also reflect the increasing willingness to bear risk as disease progresses and treatment options narrow.

'Adaptive licensing'/development

Early access to medicines schemes need to be progressed imaginatively, addressing unsolved problems, such as how prices can reflect additional value proven after initial conditional licensing.

Reimbursement decisions

Regulators and payers should take account of forms of evidence besides classic RCTs and include 'value' factors going beyond the cost/ QALY to reflect the life situation of the patients and their carers.

Point of diagnosis

Better support is needed for clinicians on giving diagnoses and developing the doctor–patient relationship. There also needs to be active encouragement

(*continued*)

Box 5.7 Continued

for clinicians to be research-active if patient willingness to participate is to be maximized.

Adherence

More research is needed into the psychological factors that lead patients (including those with life-limiting diseases) to take inappropriate self-treatment strategies and 'drug holidays'.

Health systems' uptake of innovation

The challenge is to create a health services environment conducive to truly sustainable innovation and service improvement.

---

sufferers moving into wheelchairs, there have not been adequate measures of their mobility and quality of life *after* this has happened. Trials were focused on quantitative measures of mobility while patients were still on their feet. If you are a teenager with muscular dystrophy, as one mother pointed out to me, the mere ability to move a single finger to operate a mechanized wheelchair can be a major benefit.

Clinical trials

The very word 'subjects' that is used in clinical trials underscores the problem. Patients and their organizations are too often seen as passive participants, by the innovating company and by the doctors through whom the trials are conducted. In contrast, patient organizations can advise on trial design, comment on inclusion and exclusion criteria, help recruit patients, advise on the methodology used, and assist in the education of patients—all of which will aid the success of the trial. And the individual trial patients need to be treated as active participants with the right to input and respect, and to be communicated with when the results are known.

There are many encouraging signs of greater patient engagement throughout the innovation process. As we saw in Chapter 4, the development and approval of HIV drugs were driven ahead by patient pressure. Also, the Cystic Fibrosis Foundation in the USA played a major role in the development of ivacaftor, partnering with the pharmaceutical company developing the drug and advocating its regulatory and reimbursement approval.

Patients need training to be effective in these complex, professionally dominated areas. In Europe, there is an organization called EUPATI (European Patients' Academy on Therapeutic Innovation) that trains patients in everything from research participation to adherence support. Myeloma UK has taken a very active role in helping to structure trials, recruit patients, and present 'value' and quality of life arguments to NICE.

The FDA engages patients in its advisory committees and holds patient-focused drug development sessions. They also mount a 'patient representative program' that creates a pool of screened and trained patients to take part in policy discussions. The EMA employs a somewhat different mix: periodic engagement through the drug development cycle, patient review of package leaflets and safety messages, and a permament advisory group drawn from patient organizations. The two agencies should be commended for jointly sponsoring a 'Patient Engagement Fellowship'[47]. The Fellow was able to recommend to each agency what they could learn from the other.

HTA agencies are also beginning to listen to patients. NICE routinely invites patients to its appraisal committees, as well as a Citizens' Council for debate of broader policy questions.

So the process is well underway. We are moving from zero patient input to increasing levels of representation. The next step is full partnership in decision-making. To achieve this we need not only further training of all those who participate but also clarity on when it is vital to have a current patient, versus a 'patient representative'. The latter may bring professionalism and continuity, but only the former can really inform judgments on treatment benefits and consequences.

Patient views are particularly important when we are making change to the established order. Adaptive development is a case in point. In a study for the English Department of Health, CASMI set up several patient focus groups (from late onset diseases like Parkinson's, through life-limiting adult diseases like ALS/MND to the families of children with muscular dystrophy) to discuss the implications of this change. Their responses were very revealing—see Box 5.8.

Box 5.8 Patient observations on regulatory system changes

Firstly, they described the current regulatory system as 'not working ... inflexible ... because designed by institutions ... with the primary goal of reducing risk'. Pharmaceutical companies also came in for criticism, as 'seeking to cover their own backs'. The idea of more rapid and flexible system, with a more patient-centric approach was strongly supported.

However, patients were also alert to risks and drawbacks, such as not collecting sufficient data to convincingly demonstrate whether the product works. Also that data might not be transparently available to everyone involved, including them. Patients want as much information as possible, as quickly as possible. It was also felt that many drugs are approved on the basis of endpoints that mean little or nothing to the patients themselves. There was a call for a more holistic, patient-centric way of assessing benefit.

Finally, there was concern about the potential withdrawal of a product that had been provisionally approved if the overall statistics did not bear out its efficacy, or long-term reimbursement was not forthcoming, yet an individual patient had received benefit.

It was striking that many of these points reinforced the 'Patient Manifesto', even though the two exercises were completely separate.

Professionally led and organized patient organizations are ready and eager to play their role. While we should ensure that actual patients are involved wherever feasible, the leaders of these bodies can help create the right channels for communication.

For a fairly extreme version of patient engagement we can take the story of a patient, Kim Goodsell. An athlete, diagnosed with both ventricular arrhythmia and Charcot-Marie-Tooth disease, tracked down the connection between these apparently unrelated conditions to a single mutation in one of her genes. She became her own clinical geneticist[48]. As information on genetics and the tests themselves become more widely available, more and more patients may be able to do the same.

Finally, I believe that it is the patients that should hold the key to health data integration. In the era of Big Health Data, we have two broad and disconnected realms of data about patients: (1) those generated professionally by medical practitioners, diagnostic laboratories, and other 'official' inputs into electronic medical records or research profiles, and (2) those produced by the patients themselves—either intentionally, as in patient-reported outcomes, or incidentally, in their social media interactions. We will make the greatest progress when these can be combined, ideally under the control of the patients themselves.

The golden thread of precision medicine

Precision medicine is the golden thread that runs through most of the changes advocated in this chapter, from how we define diseases to how we engage patients in their own treatment choice and wellness strategies. Precision medicine has also been called personalized medicine—or stratified medicine, for the professionals looking at patient populations. There may be subtle differences between how each could be defined but I prefer to regard them all as a single arrow of change.

Some may smile at the choice of the term 'precision medicine'. Of course, no doctor through the ages has set out to practise *imprecise* medicine. They were always seeking to make as accurate a diagnosis as possible and match the treatment to the patient. It is simply that the tools to gain specific molecular insights and to support precision or personalized therapy have multiplied dramatically in the last 20 years. And they look set to expand even more rapidly in the next 20. However, precision medicine, as it is now being defined and equipped, is not just a set of insightful tools and valuable technologies but also a powerful idea.

Let us trace the thread through the seven directions of change set out in this chapter. Firstly, a more precise diagnosis of the underlying molecular disorder that underlies symptoms leads to a more precise intervention in the pathways and systems involved. Some of these findings will emerge from new academic–industry collaborations. With this stronger scientific base, we will see more precisely targeted clinical candidates. More of these potential therapies will emerge through trials as approved and reimbursed products, since patients will be more precisely chosen and outcomes more precisely measured.

Box 5.9 Precision medicine and the gaps in translation

T0: More precise understanding of disease and how it should be sub-segmented will ensure that subsequent discovery efforts are well targeted and have markers of disease and response built in.

T1: Precision medicine tools will lead to research that is more translatable from laboratory into candidate products, albeit for smaller groups of patients. Markers and models will reflect a more precise disease definition at the molecular level.

T2: Clinical trial subjects will be selected using the right markers. Regulators and reimbursement agencies will be ready to accept data earlier on smaller but more responsive groups, since it will be more persuasive, and budget impacts easier to quantify.

T3: Doctors will have specific tests to run to check their patients for suitability. Patients will have more confidence in the fit of the medicine to their needs. Personal tools for adherence support and monitoring will assist patients to get the full benefit.

T4: Mining the accumulated health data will result in findings that extend further our understanding of real world response and safety performance, and give us insights into sub-groups with either better or worse benefit/risk.

Doctors will be more confident in their use and will know the precise tests that should be run before a treatment choice is made. Patients, approached as individuals, will be more proactive in agreeing a more personal, and therefore more precise therapy plan, and more receptive to adhering to it. Such a plan will centre on the patient, spot the emergence of a disease or track its course, and monitor the response to treatment, using precision digital tools shared between patient and clinician. Patient engagement throughout the innovation process will keep the professionals focused on what benefits, risks, and values match patients' situations most precisely.

Pardon the repetition of 'precise', but I believe precision medicine is the theme that connects the seven separate elements in our change programme. This also helps get non-professionals, like politicians and the media, behind the changes they will need to support, as we will discuss in Chapter 6.

We can also trace the relevance to each of the 'T' gaps in translation, as we build 'precision medicine solutions' for each of them—see Box 5.9.

This movement towards much more personalized, more precise medicine will occur in three main phases. But it is useful to summarize where we start.

Current practice

Our starting point is the current approach, which combines differential diagnosis and 'trial and error'. Differential diagnosis seeks to home in on the most accurate explanation for the presenting problem, using the full set of symptoms, patient and family history, and accumulated medical knowledge. Probability plays a major role in the old medical saying: 'When you hear hoof beats, look for horses, not zebras'—in other words expect the most common diagnosis but rule out more exotic and especially more dangerous conditions as quickly as possible.

For non-medical professionals, a recent personal example: I had a painful swelling in my foot: did I bruise it on something (most likely), was I stung by something (possible), do I have bursitis, or cellulitis, or might it be deep vein thrombosis (most dangerous)? Only an ultrasound can distinguish in such cases. Treatments would be very different, depending on the diagnosis.

Today, once a differential diagnosis is made, doctors recognize that not all patients will respond equally to the most common treatment choice. 'Trial and error' is an unkind but fundamentally accurate description of much prescribing practice, in areas in which diseases may take a variety of forms. Licensed drugs have a variety of attributes and therefore patients a variety of responses: 'Let me start you on this and see how you do.'

The current approach can, of course, be made more scientific, through algorithms that ensure that the clinician draws upon the accumulated experience of the field. Some can become very sophisticated, as for example the NCCN guidelines for cancer, many of which span several pages for a single disease (and are increasingly populated with some of the precision medicine insights we now go on to describe), individualized and predictive, as depicted in Figure 5.5.

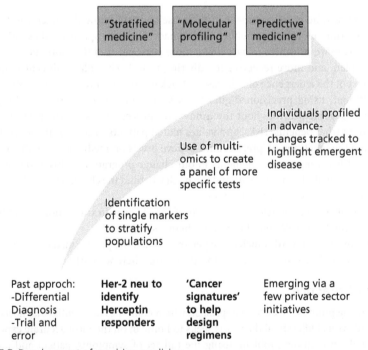

Fig. 5.5 Development of precision medicine.

Stratified medicine

'Stratified medicine' uses a single biomarker to divide disease populations into groups (typically two—with and without the biomarker) with major implications for treatment choice.[49]

The rationale and evidence base for much precision or stratified medicine is fairly straightforward: the proposed treatment targets a specific cellular component identified by a diagnostic test, whether for the presence of a gene or protein. In the ideal case a specific mutation corresponds to one or more receptor proteins that are: (1) targetable by a drug, (2) have a major impact on the development of the disease, for example via a signalling pathway, and (3) are on the critical path to modifying the behaviour of the cell, in a way that is not immediately compensated for. Many diseases, of course, are not as simple as this, and are underpinned by complex pattern of interactions rather than a single molecular error.

Discrete mutations are the cause of many cancers. The 'poster child' for this phase of precision medicine has been Herceptin, a treatment that targets a specific antigen on the surface of a breast cancer cell. The presence or absence of the 'Her-2-neu' receptor is established with a diagnostic test, already commonly run on breast cancer patients in many parts of the world. This test has the benefit of being readily understandable by the clinician and even the patient: if your tumour cells have the receptor then the drug is right for you.

As is often the case in cancer, Herceptin was first introduced in late stage, metastatic patients and then found to be a valuable part of earlier stage treatment, with excellent impact on disease progress and survival.

In cancer, this diagnostic/drug combination was among the first of several others. For example, mutations in the *KRAS* gene (first discovered in mouse sarcomas) allow a protein p21ras to be produced, that is on a pathway that sends cell growth signals to the tumour cell's nucleus. The seven most common *KRAS* mutations are found in virtually all colorectal cancers, and the *KRAS* status predicts response to therapies such as panitumumab.

Regulators have responded to these new drug/diagnostic combinations. In the last few years, there has been a major increase in the number of precision or personalized treatments approved, from 13 in 2006 to 113 in 2014. The regulators have tried to keep pace with the science, with ten separate policy and guidelines documents issued by the FDA over the same period, ranging from guidance on 'pharmacogenomic tests' to co-development of such tests and the drugs themselves.

In contrast, *reimbursement* has lagged. In the early days, tests were reimbursed in the USA via 'stacked codes': adding together the reimbursement associated with the various steps or elements in the test. Changes to US payments systems via insurance contractors, however, inadvertently brought federal reimbursement to a standstill by 2013. The situation in Europe is little better.

In some cases, even diagnostics that identify patients that are unlikely to need expensive chemotherapy at all have not been reimbursed by the systems that stand to save money. Oncotype Dx (from Genomic Health)—a test that determines if a woman will benefit from post-operative chemotherapy—has taken several years to pass NICE and to begin to be reimbursed in the UK.

Precision medicine is now making its appearance in other areas, such as several of those reviewed in Chapter 4. Neonatal diabetes and cystic fibrosis were two conditions we saw in which different genetic factors produce similar symptoms. In these cases, there is again a match between a molecular target and a therapeutic approach.

As well as opening the door to improved efficacy, precision medicine is also the key to minimizing the risk of side-effects. Many of the most common drugs are metabolized by two enzymes in the liver—cytochromes CYP2D6 and CYP2C19. Mutations in the corresponding genes, resulting in different activity levels for one or other of these cytochromes, explains the wide variety of drug responses found in the population. There are essentially four types of metabolic situations—from ultra-rapid metabolism (in which the drug may be broken down too fast to have a sustained therapeutic effect) to poor metabolism (in which the drug stays un-metabolized in the system for an extended period and causes side-effects).

This is one of the complex underlying reasons for the estimated 40–60% of patients that do not benefit from commonly used drugs[50]. Table 5.1 contains estimates of such failure rates in different disease areas.

Table 5.1 Drug efficacy rates[a]

Therapeutic area	Efficacy rate (%)
Alzheimer's	30
Analgesics	80
Asthma	60
Cardiac arrhythmias	60
Depression (SSRI)	62
Diabetes	57
HCV	47
Incontinence	40
Migraine	52
Oncology	25
Osteoporosis	48
Rheumatoid arthritis	48
Schizophrenia	60

[a] Another author has dug into this much-cited study and believes that a proportion of the reported non-response derives from stochastic variations in the responses of individuals (see 137).

Reprinted from *Trends in Molecular Medicine*, Volume 7, Issue 5, Spear BB, Heath-Chiozzi M, and Huff J, Clinical application of pharmacogenetics, pp. 201–204, Copyright © 2001, with permission from Elsevier.

Source: data from Medical Economics. (1999) *Physicians' Desk Reference*, 54th edition. Medical Economics Company, Inc.

The benefits for the patient and health system from biomarker-directed therapy are clear—the treatment is only used when the molecular marker indicates it will be effective. However, the economics for the innovators can be problematic, both at the development stage and once on the market. The widespread commercial availability of the diagnostic is critical for the success of the approach, but the kind of clinical trial validation needed is beyond the resources of diagnostic companies. So the pharmaceutical partner must typically sponsor the trials. When the combination is licensed the costs of the diagnostic are typically higher than most diagnostic budgets can afford and the test may need to be subsidized or even given away alongside the drug.

In time, of course, other pharmaceuticals are developed with the same molecular target, which confuses the economics further, although of course it expands the market for the diagnostic. If the first drug innovator seeks to control the companion diagnostic market, this obviously limits competition, to an extent that is undesirable from the payers' perspective.

Molecular profiling

As we develop our insights into the personalized nature of some diseases further, we enter the next stage, which I term 'molecular profiling.' This recognizes that, in many diseases, a range of molecular characteristics can mark variation between patients. Several genes and multiple proteins may differentiate their disease at a cellular, organ, or organism level.

These molecular signatures have been mapped out with increasing sophistication in cancer since chips were developed to measure the variety of 'transcripts' in tumours that seemed similar in classic pathological examination. This approach has now moved to the genomic level with the development of genomic signatures that sub-segment disease, as laid out in a library of signatures recently published by the UK's Sanger Centre[51].

Likewise, panels of proteomic tests have been shown to differentiate sepsis patients. Sepsis is very broad description of progressive failure of the immune system in the face of systemic infection, often leading to sequential organ failure and death. It has defeated the efforts of many drug developers, despite the size and importance of the potential market, in part because of its heterogeneity. The Yale spin-off company Molecular Staging that I once led discovered four specific serum proteins that distinguished those sepsis patients likely to be terminal and those most responsive to the only drug available at the time, Xigris.

The validation of these kinds of signatures is more complex than for single biomarkers. On the one hand, adding to the list of markers could result in more accurate definition of the disease; on the other, the more markers that are included, the higher the chances of false correlations. In any case, such signatures are only usable clinically when they have been demonstrated to affect clinical outcomes.

These diagnostic panels are most likely to emerge from academic research, since they are not linked to a single drug. We will need an economic basis for their validation and deployment, perhaps via 'pre-competitive' consortia of companies. We have seen the beginnings of this approach in the work of the Critical Path Institute on Alzheimer's and Parkinson's diseases. Various potential markers in blood and cerebrospinal fluid and imaging have been developed and accepted by the FDA to sub-segment patients for future clinical trials[52].

The immune system is likely to be a fertile field for molecular profiling, as there are multiple types of cells and cytokines to track. Their abundance or nature indicates either the health of the system, its response to infection or its role in auto-immune diseases. This 'immunophenotyping' is a field of active research. The Human Immunology Project, run by the National Institute of Health's Division of Allergy, Immunology and Transplantation, has set out to 'characterize various states of the human immune system following infection'. (When fears of bioterrorism were at their height, there were various efforts to identify rogue organisms by this approach.) Such a broad, openly available resource has obvious value for vaccine development as well as for the design of drugs to modify (up- or down-regulate) immune system activity.

What will be needed to take molecular profiling from useful research tools to routine clinical use? In short, a viable business model. Consortia can fund their development, probably on an 'open source' basis, but then companies will be needed to commercialize the genomic or proteomic panels on user-convenient instruments, a costly investment.

Predictive medicine

The ultimate stage of precision medicine goes beyond distinguishing groups of patients to tracking the *emergence of disease and treatment response through time in single individuals*. This can be termed 'predictive medicine' in that it can: (1) define the predisposition of an individual patient to a specific dizease, (2) identify its emergence, and (3) track its progress in the face of various treatment regimens.

Again cancer will probably be the testing ground. We are beginning to map the evolution of cancer cells within tumours. Sequential mutations give rise to different colonies of cells, and these are differentially affected by different regimens of drugs, and we will be able to track this process. Also, as we seek to manipulate the immune environment, immuno-phenotyping can play a role. So both the tumour and its environment can be tracked as treatment progresses.

In time this truly personalized, multi-omic approach will benefit patients with inflammatory diseases, developmental and metabolic diseases, and sepsis. The potential is exciting, but the scientific, developmental, and economic challenges are all profound. But early work in this direction is promising. As we saw, Mike Snyder at Stanford tracked a single individual's multi-omic profile through time in the face of life events, such as infection, and monitored their impact on health.

Lee Hood's work at the Institute of Systems Biology on 107 healthy individuals is another example, in this case combining 'multi-omic testing with health coaching'[53]. This is being launched as a commercial 'scientific wellness' company, Arivale, that describes itself as: 'combining cutting edge science, an intimate and unprecedented view of your body, and personalized coaching to help you achieve your unique wellness potential.' Another great genomics pioneer, Craig Venter, has created Human Longevity Inc, with a similar ambition, but a somewhat different set of tools.

It may be some time before such programmes become available to the general population and I can already hear the objections from medical experts that believe we are already 'over-medicalizing life'. Ultimately it will be the consumers who decide.

National precision medicine programmes

As precision medicine emerges as a discipline and strategic direction, various countries are making substantial investments. Britain has invested more than a billion pounds in precision medicine-related research programmes in recent years. It created the UK BioBank, with its 500 000 individuals whose samples and key medical information are centralized and searchable. The Medical Research Council has created defined patient cohorts in several diseases, and laid out a national network of molecular pathology labs. The UK's '100 000 Genome' project is sequencing the genomes of patients with inherited diseases and their parents, and patients with the common cancers. To complement such research programmes, new databases are bringing together the anonymized medical record information of millions of people. The UK Precision Medicine Catapult is facilitating the creation and growth of businesses to exploit these assets. If these can be brought together in a connected strategy, the potential is enormous.

In France, the French National Cancer Institute has established a network of 28 molecular genetics centres to exploit the power of molecular diagnostics to target cancer therapy.

This is one of the few areas in which Europe may be leading the USA, at least in terms of national (i.e.federal) initiatives. President Obama, working with the head of the NIH Frances Collins, has launched the USA's own precision medicine initiative, one component of which is to build a US equivalent of the BioBank (but, inevitably, one aiming to be much larger!).

Precision medicine may initially move fastest in advanced parts of the US system, like the hospitals in the University of California group, bringing together their separate plans and efforts. In a recent paper 'Precision Medicine—Beyond the Inflection Point', the Chancellor of UCSF, Sam Hawgood, described precision medicine as a new era in medicine: 'A confluence of biological, physical, engineering, computer and health sciences is setting the stage for a transformative leap toward data-driven, mechanism-based health and healthcare for each individual'[54]. Ambitious words, setting a bold agenda, very much aligned to the theme of this book.

What might a 'precision medicine strategy' look like? Figure 5.6 shows some of its key elements, both in terms of R&D and subsequent medical practice. The feedback loop from RWD to disease definition recognizes the fact that precision medicine will be progressively learning from its own application.

As precision medicine advances, it is tempting to focus only on the biological aspects of personalization. But we must not overlook two other key dimensions on which we can personalize medicine: the psychological and the situational. Individual health behaviour—resulting from different self-image, beliefs, and concerns—clearly shapes medical needs. Likewise, medicine should not operate in a social vacuum, without factoring in the support and other influences coming from the family and community

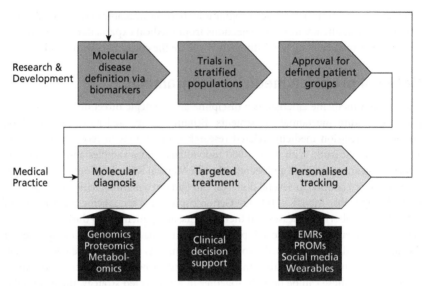

Fig. 5.6 Precision medicine schematic.

situation in which patients find themselves. As precision medicine proceeds we will need to bring together all of these three dimensions of personalization, in the strategy for each individual patient.

The final, and perhaps toughest challenge, is how precision or personalized medicine is to be paid for. The markets will be, by definition, smaller than those that innovators have traditionally addressed. What incentives can we put in place for them to follow the path that science lights up for the future?

The first incentive resides in the shorter and smaller clinical trials that precision medicines will need. The second, and related, incentive follows from the adaptive development approach of which precision medicine is an integral part. If we allow precision medicines on to the market earlier, because of their more precise targeting, this has competitive and financial advantages for the innovator.

But we may need to think more radically. Firstly to be prepared to attach higher prices to these smaller markets, perhaps linked to the outcomes achieved. Payers should be more receptive to this if they know their total exposure is limited, and they may be prepared to do a total volume deal. In such a deal the innovator would see no incentive to promote or allow the product to be used more widely than defined by the 'precision' label.

Even more fundamentally, the era of precision medicine may prompt us to look again at how we reward innovators via market exclusivity. Today this exclusivity is defined as the product's patent life plus a period of 'data exclusivity'. (In the latter period potential generic competitors cannot rely on—or refer to—the clinical trial data submitted by the original innovator.) While this time-limited monopoly is a vital source of incentive for the innovator, it is a rather blunt instrument and links the incentive to the original discovery step (which may not have involved much investment) rather than to the development process (that always does).

This model does not fit a world in which innovators explore and launch in smaller, niche markets and then add to them over time. Currently the exclusivity period can run out before the innovator has explored all the possible indications. Perhaps we should develop a formula to link the period of exclusivity to the size of the market. We have taken the first steps in this direction with orphan drug legislation on both sides of the Atlantic. A similar approach could allow a longer exclusivity period for an innovator who volunteers to restrict the initial use of a drug to a small sub-segment of the patient population and then plans to trial in other markets later.

The other half of the equation is, of course, the pricing and affordability question we touched on earlier. Health systems and payers need to be persuaded that what they pay, in terms of high prices for precisely targeted medicines, are outweighed by the savings from otherwise untargeted and therefore often ineffective prescriptions.

Precision medicine is not a panacea, and raises new and important issues. Some of these we will explore in Chapter 6, on the public and policy environment, which is currently largely unprepared for the precision medicine revolution.

Summary

Pulling together all seven steps to sustainability, we see how they form a radically new model of innovation, tackling all our gaps in translation, with precision medicine a theme throughout (Figure 5.7).

The relative importance and impact of these steps will vary from one disease to another. In some diseases, like ALS, we need the most effort to overcome T0 and T1, to uncover more about the molecular basis of disease, find good disease models, and generate early leads for preclinical testing. In cancer, with a large number of promising products already in the pipeline, T2 looks to be the priority. We need to ensure that well-targeted products receive a rapid passage through approval and that reimbursement formulae are found that do not inhibit patient access.

In major chronic diseases for which we have effective, underutilized therapies, for example asthma, we may need to focus on T3, targeting treatments effectively to sub-types. We must also use electronic tools, based on our insights into patient psychology, to boost adherence.

Closing the T4 gap will be of greatest relevance for diseases of high prevalence, as we study the impact of therapies over time, spot important drug–drug interactions, and highlight positive and negative lifestyle factors.

These are my personal sense of the priorities: it is the research community that should debate them, so that funding and effort is placed in the areas of highest payback.

If this seven-point plan to precision medicine is the right strategy to close the innovation gap, whose job is it to implement it? It is no one organization's role to think and act at this strategic level. But research funders and healthcare payers, particularly major governments, should embrace this perspective. They can and should ensure the signals they send, in the form of research grants, regulatory and reimbursement decisions, and population health initiatives, 'nudge' the field in the right direction.

This brings us to a final point that applies across the whole innovation process. Successful innovation has to be approached from *both* a left-to-right and right-to-left

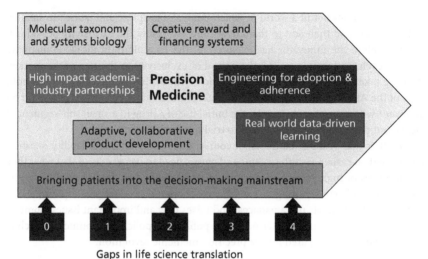

Gaps in life science translation

Fig. 5.7 Seven steps to sustainable innovation.

direction. Most innovators start with a technology and ask, often sequentially over a period of years, the key technology, market and economic questions: (1) Does the technology work? Can I turn it into a new product or service? (2) Where and how can I apply it—where are the markets? (3) Will someone pay for it, and why?

This is what I call left-to-right analysis. This kind of sequential thinking is understandable, but unfortunately exciting technology often falls at the market or economic hurdles, for reasons that might have been predicted from the outset.

To avoid this, left-to-right analysis must be complemented by right-to-left thinking, starting from the perspective of the ultimate customers:

◆ **Need:** Where are health systems' most important problems, from the clinical and economic standpoint?

◆ **Market:** What kinds of solutions will the market accept and will payers reward? And what kinds of prices are they likely to be acceptable?

◆ **Patients:** Will patients recognize the benefits, embrace the products, and adhere to their use?

◆ **Clinicians:** Will clinicians adopt the technology or resist it, as a result of skills gaps or vested interests?

In other words, does this technology address genuine health system priorities in a way that patients, clinicians, and payers will recognize?

Scientists and technologists naturally work forward from their discoveries. Entrepreneurs are skilled at connecting with market need and economic viability. The best innovative companies contain both top-flight scientists and savvy entrepreneurs. And the wisest payers and governments devise mechanisms for dialogue that ensure those entrepreneurial companies know where the greatest health rewards will lie, and pursue them.

Summary of Chapter 5

Bioscience itself is producing tools that can streamline and focus the medical innovation process. Drawing on these tools we must make seven specific changes of mindset and practice. Together they have the potential to transform life science translation, from a new approach to discovery, through faster, more targeted therapy development, to a better use of health data. Together these changes spell a new era of precision medicine.

In Chapter 6 we examine the environment—political, public, media, ethical, and financial—in which these changes need to occur. Here too we need new thinking and receptivity to change, a new set of policy goals and a new quality of public dialogue.

References

1. **Patricia Seemann,** website of the '3am Group', http://www.the3amgroup.org, accessed 01 Sep. 2015

2. **Chen R, Mias GI, Li-Pook-Than J,** et al. 2012. Personal omics profiling reveals dynamic molecular and medical phenotypes. *Cell,* 148:1293–1307.

3. **Nisbett RE,** *The Geography of Thought: How Asians and Westerners Think Differently...and Why* New York: Free Press.

4. **Alexander Gelfand,** Building a full-blown human body-on-a-chip, 30 April 2015, http://discovermagazine.com/2015/june/4-pieces-of-me, accessed 01 Sep. 2015

5. In Silico Biology, journal published by STM Publishing House, http://www.iospress.nl/journal/in-silico-biology/, accessed 01 Sep. 2015

6. **Thal LJ, Kantarci K, Reiman EM.** 2006. The role of biomarkers in clinical trials for Alzheimer disease. *Alzheimer Dis Assoc Disord,* 20: 6–15.

7. **Bakkar N, Boehringer A, Bowser R.** 2015. Use of biomarkers in ALS drug development and clinical trials. *Brain Res,* 1607:94–107

8. **Kelloff GJ, Sigman, CC.** 2012. Cancer biomarkers: selecting the right drug for the right patient. *Nat Rev Drug Discov,* 11: 201–14.

9. **Kim D, Labkoff S, Holliday SH.** 2008. Opportunities for electronic health record data to support business functions in the pharmaceutical industry—A case study from Pfizer, Inc. *J Am Med Inform Assoc,* 15:581–4.

10. **Uren SC, Kirkman MB, Dalton BS, Zalcberg JR.** 2013. Reducing clinical trial monitoring resource allocation and costs through remote access to electronic medical records. *J Oncol Pract,* 9:e13–6.

11. **Beyer ARH, Fasolo B, de Graeff PA, Hillege, HL.** 2015. Risk attitudes and personality traits predict perceptions of benefits and risks for medicinal products: a field study of European medical assessors. *Value Health,* 18:91–9.

12. **McCrae RR, Costa PT.** 1987. Validation of the five-factor model of personality across instruments and observers. *J Person Soc Psychol,* 52:81–90

13. Developing an aetiological based taxonomy of disease, IMI indicative call topic, http://www.imi.europa.eu/webfm_send/565, accessed 01 Sep. 2015

14. Toward precision medicine: Building a knowledge network for biomedical research and a new taxonomy of disease, published by the National Academies Press, 2011, http://www.ncbi.nlm.nih.gov/pubmed/22536618, accessed 01 Sep. 2015

15. Committee on a Framework for Developing a New Taxonomy of Disease. 2011. *Toward Precision Medicine* Washington DC: National Academies Press.

16. **Pao W, Girard N.** 2011. New driver mutations in non-small-cell lung cancer. *Lancet Oncol,* 12:175–80.

17. **Pascoe S, Locantore N, Dransfield MT, Barnes NC, Pavord ID.** 2015. Blood eosinophil counts, exacerbations, and response to the addition of inhaled fluticasone furoate to vilanterol in patients with chronic obstructive pulmonary disease: a secondary analysis of data from two parallel randomised controlled trials. *Lancet Respir Med,* 3:435–42.

18. **Wessels JA, Huizinga TW, Guchelaar HJ.** 2008. Recent insights in the pharmacological actions of methotrexate in the treatment of rheumatoid arthritis. *Rheumatology* 47: 249–55.

19. **Böhm I.** 2004. Increased peripheral blood B-cells expressing the CD5 molecules in association to autoantibodies in patients with lupus erythematosus and evidence to selectively down-modulate them. *Biomed & Pharmacother,* 58:338–43.

20. **Brody M, Böhm I, Bauer R.** 1993. Mechanism of action of methotrexate: experimental evidence that methotrexate blocks the binding of interleukin 1 beta to the interleukin 1 receptor on target cells. *Eur J Clin Chem Clin Biochem,* 31:667–74.

21. **Pigott R, Barker R, Kaan T, Roberts M.** 2014. Shaping the Future of Open Innovation, http://www.wellcome.ac.uk/About-us/Publications/Reports/Biomedical-science/WTP057246.htm).

22. **Barker R.** 2010. A flexible blueprint for the future of drug development. *Lancet,* 375: 357–9.

23. Adaptive Pathways, European Medicines Agency guidance document, http://www.ema.europa.eu/ema/index.jsp?curl=pages/regulation/general/general_content_000601.jsp, accessed 01 Sep. 2015

24. **Barker R, Garner S.** 2015. Realising the potential of adaptive development. *Rev Recent Clin Trials,* 10:19–24.

25. **Baird LG, Banken R, Eichler HG,** et al. 2014. Accelerated access to innovative medicines for patients in need. *Clin Pharmacol Ther,* 96:559–71.

26. Pilots of multi-stakeholder consultations in drug development, 6 June 2012, http://www.tapestrynetworks.com/initiatives/healthcare/upload/Pilots-of-multi-stakeholder-consultations-in-drug-development-6-June-2012.pdf, accessed 01 Sep. 2015

27. Medicines Adaptive Pathways for Patients (MAPPs), report by CASMI for the UK Office of Life Sciences, https://www.gov.uk/government/uploads/system/uploads/attachment_data/file/509647/mapps.pdf, accessed 01 Sep. 2015

28. **Carbonneil, C, Quentin F, Lee-Robin SH,** European Network for Health Technology Assessment. 2009. A common policy framework for evidence generation on promising health technologies. *Int J Technol Assess Health Care,* 25.S2: 56–67)).

29. **Rawlins M.** 2008. De testimonio: on the evidence for decisions about the use of therapeutic interventions. *Lancet,* 372:2152–61.

30. The I-SPY trials: Personalized medicine and novel clinical trial design http://ispy2.org, accessed 01 Sep. 2015

31. Revolutionary clinical trial aims to advance lung cancer treatment, media release on Cancer Research UK website, 17 April 2014, http://www.cancerresearchuk.org/about-us/

cancer-news/press-release/2014-04-17-revolutionary-clinical-trial-aims-to-advance-lung-cancer-treatment, accessed 01 Sep. 2015

32. **Menis J, Hasan B, Besse B.** 2014. New clinical research strategies in thoracic oncology: clinical trial design, adaptive, basket and umbrella trials, new end-points and new evaluations of response. *Eur Respir Rev*, 23:367–78.

33. **Barker RW.** 2011. *2030 The Future of Medicine—Avoiding a Medical Meltdown.* Oxford: Oxford University Press.

34. **Jaczynska E, Outterson K, Mestre-Ferrandiz J,** February 2015, Business model options for antibiotics; Learning from other industries, Joint publication of Chatham House and Big Innovation Centre, http://drive-ab.eu/wp-content/uploads/2014/09/Business-Model-Options-for-Antibiotics-learning-from-other-industries.pdf, accessed 01 Sep. 2015

35. Which doctors take up promising ideas? Report by CASMI for Nesta, 28 Jan. 2014 https://www.nesta.org.uk/sites/.../which_doctors_take_up_promising.pdf, accessed 01 Sep. 2015

36. Diabetes Improvement Collaborative 2014–2015, report from the Health Innovation Network on AHSN Network website http://www.ahsnnetwork.com/wp-content/uploads/2014/12/HIN-Diabetes-Improvement-Collaborative-Case-study-2.pdf, accessed 01 Sep. 2015

37. Right insulin, right time, right dose Report by Health Innovation Network on website hinsouthlondon.org, accessed 01 Sep. 2015

38. **Horne R, Chapman SCE, Parham R, Freemantle N, Forbes A, Cooper V.** 2013. Understanding patients' adherence-related beliefs about medicines prescribed for long term conditions: a meta-analytic review of the Necessity-Concerns Framework. *PLoS ONE*, 8:e80633

39. **Petrie KJ, Weinman S.** 2012. Patients' perceptions of their illness. *Curr Dir Psychol Sci*, 21, 60–5.

40. Website of the Cochrane Collaborative, www.uk.cochrane.org, accessed 01 Sep. 2015

41. **McGuiness B** et al; Cochrane Library; 2008

42. **Ho C, Murray N, Laskin J, Melosky B, Anderson H, Bebb G.** 2005. Asian ethnicity and adenocarcinoma histology continues to predict response to gefitinib in patients treated for advanced non-small cell carcinoma of the lung in North America. *Lung Cancer*, 49:225–31.

43. **Kasznicki J, Sliwinska A, Drzewoski J.** 2014. Metformin in cancer therapy and prevention. *Ann Transl Med*, 2:57

44. IMI GetReal project, reports published on IMI website, http://www.imi-getreal.eu, accessed 01 Sep. 2015

45. *Fortune Magazine*, July 30, 2015

46. The Patient Manifesto, published on the CASMI website, www.casmi.org.uk/the-patient-manifesto/, accessed 01 Sep. 2015

47. Blogs.fda.gov; February 12, 2015.

48. The amateur geneticist who surprised science, report on BBC website of work by Kim Goodsell, http://www.bbc.com/future/story/20140819-the-amateur-who-surprised-science, accessed 01 Sep. 2015

49. **Spear BB, Heath-Chiozzi M, Huff J.** 2001. Clinical application of pharmacogenetics. *Trends Mol Med*, 7:201–4.

50. Responder despondency: myths of personalized medicine, Guest Post by Stephen Senn, on Error Statistics Philosophy website, http://errorstatistics.files.wordpress.com/2014/07/senn-responder-table-1.png, accessed 01 Sep. 2015

51. COSMIC: Signatures of Mutational Processes in Human Cancer, published on Sanger website http://cancer.sanger.ac.uk/cosmic/signatures, accessed 01 Sep. 2015

52. Critical Path Institute secures regulatory support for Parkinson's and Alzheimer's disease biomarkers, published on Critical Path Institute website http://c-path.org/critical-path-institute-secures-regulatory-support-for-parkinsons-and-alzheimers-disease-biomarkers%E2%80%8B/, accessed 01 Sep. 2015

53. Leroy Hood describes promising results from pilot stage of ISB's 100k wellness project https://www.genomeweb.com/molecular-diagnostics/leroy-hood-describes-promising-results-pilot-stage-isbs-100k-wellness-project, accessed 01 Sep. 2015

54. **Hawgood S, Hook-Barnard IG, O'Brien TC, Yamamoto KR**. 2015. Precision medicine: Beyond the inflection point. *Sci Transl Med*, 7(300) 300ps17.

Chapter 6

Getting the environment right

The environment for medical innovation, and the support for the changes we need to make, is shaped by the opinions, policies, and actions of many external stakeholders. Prominent among them are politicians, the media, the general public, and, most importantly, the patients themselves. No plan to revolutionize medical innovation would be complete without a strategy to engage and align these opinions.

Since the field, and some of the changes, are rife with genuine ethical issues, we must also invest in objective ethical study. It is vital the changes we propose are understood widely as in the interest of patients, rather than appearing designed to make life easier for innovators. Health data, its ownership, access, and utilization, is a particular environmental issue we need to get right.

We also need the development of the field of regulatory science, to provide scientific underpinning for many of the changes described in the Chapter 5. And we need to encourage the efforts of public–private partnerships and consortia that set out to develop solutions of value across the whole life science enterprise. We also need to understand and encourage the building of life sciences clusters, in which most of the advances and their early implementation occurs.

A crucial part of the environment, of course, is the availability of funding at each stage of development. The current funding environment is quite positive, in some parts of the world. However, we need to plug some gaps here also, or the resources will not be there to bridge translational gaps and accelerate innovation.

Most fundamentally, we need to examine and reform society's ingrained attitudes as they relate to the process of medical innovation. We need nothing less than a new 'social contract' for medical innovation—since the new science and the new models will both challenge what people currently believe and expect.

Politics

Political opinion is important on several dimensions: R&D funding decisions, direction given to regulatory bodies, health system policy, and reimbursement priorities.

Publically funded R&D has always had a crucial role in life sciences. Apart from a few major foundations, such as the Howard Hughes Foundation, the Bill & Melinda Gates Foundation, and the Wellcome Trust, and some specific disease-focused charities, the funding for basic research is essentially the responsibility of governments. Without this seed corn, the rest of the innovation process is barren.

Life sciences receive a major share of public research investment. Political whims and fashions affect the appetite to continue to support life sciences. However, senior politicians who have seen health problems in their own families will continue to be receptive to the argument that bioscience research is essential if we are to find cures for unconquered diseases. The doubling of the US NIH budget in the 1990s is a case in point. But finance ministries need to see returns, in the form of successful businesses or health system efficiencies, or they will propose cuts in straightened times.

Most funding goes into basic research, and rightly so. However, as Mariana Mazzacatto has pointed out[1], governments can and should also invest in applied and developmental programmes. This is particularly important in areas that look too novel or risky to yet attract the venture capital industry (which includes many hot areas in life sciences).

In the USA, some of the budgets of the NIH as well as military programmes sponsored by the DOD laboratories or DARPA have a very important stimulatory role in bridging the T1 gap. The German government, with its Fraunhofer Institutes, and the French CNRS invest along similar lines. Many of the most recent UK programmes follow the same early development path, including the creation of cell therapy, medicines technology, and precision medicine 'catapults' by InnovateUK, and industry collaboration programmes on the part of the Medical Research Council.

There is much debate about where public sector funding should end and private sector investment begin. Governments are urged not to 'pick winners' by supporting specific industry sectors. But it is obvious that there is still a 'valley of death' in funding many life science technologies. The consequence of any country failing to bridge this gap, either with co-funding or tax allowances, is to see those technologies flourish elsewhere. But the argument has often to be made to every new administration without senior experience in the sector. We will return to other sources of finance later in this chapter.

While regulatory bodies (such as the FDA and EMA) are typically set up at arm's length to government, there is no doubt that the latter exert huge influence on the 'solution space' in which regulators operate. Congressional influence over the FDA is strong, and has historically sometimes been pro-innovation and sometimes more 'safety-first'. In the case of Europe, it is important to remember that the EMA merely recommends approval of new products to the European Commission. An unsympathetic Commission can effectively constrain progress towards more adaptive regulatory processes.

In Europe, the adaptive pathway approach—which I believe to be central to speeding development and improving affordability—is more popular with some EU governments than others. The UK, France, Scandinavia, and the Netherlands are amongst the enthusiasts; Germany is a leading sceptic, believing that it leads to less rigorous information on new products.

Health system policy—especially when healthcare reform is on the agenda—can have a profound effect on the uptake of innovation. The ongoing sense of financial crisis in some of the European health systems, often labelled as inevitable consequences of post-2008 'austerity', can be very antagonistic towards adoption of the new and usually more expensive.

Fundamentally, when it comes to reimbursement and adoption policy for medical innovations, governments come in two kinds—those who see their country as participating actively in medical innovation and those who simply want to get the best deal on innovations developed elsewhere. Even 'participant' governments, however, often exhibit complete policy schizophrenia. Their industry or science ministries will devise initiatives to promote life sciences R&D, by supporting university research, building clinical trial capability, helping launch early stage companies, and wooing major international companies to invest. Meanwhile the health ministries, faced with the budget costs of innovations, are devising ways to control prices, constrain uptake, or both. Finance ministries, like the UK Treasury, mirror both facets in their own organizations—groups responsible for economic growth versus those controlling public expenditure.

A major tool in governmental cost control is obviously the HTA process. These are intended to ensure that choices about specific products are not politically influenced, and cannot be politically challenged. At the end of the day, however, it is governments who give them their goals and parameters.

Pharmaceutical prices are typically the sharp end of the healthcare cost control issue. For non-US payers the fact that the dominant US market has so far been ready to pay high prices means that they are frequently facing requests for prices that they see as unaffordable. HTA evaluation is used to bargain them down, but in some cases (illustrated by the recently launched HCV therapies) the drugs are cost-effective, but present a major short-term budgetary burden. Tendering arrangements are increasingly being used where there are alternative products.

US payers, public and private, complain that the US market is bearing an unfair share of the global cost of pharmaceutical R&D. Federal intervention in drug pricing on the basis of cost-effectiveness is forbidden by law; however, evaluation of comparative effectiveness is not. We will therefore see more of the latter. Where there are clear options we see competitive tendering in the USA also.

The tension between encouraging innovation and controlling prices does not seem to be seen very differently by left- versus right-leaning governments, based on my observations, on both sides of the Atlantic. Introduction of tough cost controls seem to be more a result of short-term budget crises than of political inclination.

In the UK at present, both economic growth and drug expenditure control responsibilities have been brought together under a single Minister for Life Sciences. This minister is responsible for supporting the life science sector, guiding the regulator and NICE in their decision-making, and for running the system of product pricing for NHS use. This important political experiment began only in 2014: it will be interesting to track its progress. I believe it is better than pitting ministers and officials charged with industrial development against those managing health budgets.

In the USA, bipartisan concern about the time and cost demands of new treatments has triggered the 21st Century Cures Act, being shaped in 2015 in Congress. From a quite revolutionary initial conception, it has become somewhat less ambitious but still contains important changes that seem acceptable to regulators and industry. Examples include the gathering of patient experience data for the assessment of benefit/risk, the use of adaptive trial designs and Bayesian statistics to streamline development, and the greater use of RWD. The last of these will be greatly facilitated by the Sentinel system, originally a Harvard Pilgrim innovation but since adopted and funded by the FDA.

Its common data model and ability to collect 19 key metrics consistently makes it a nationally relevant system.

The tension between supporting innovation and containing health costs will always be with us. The more successful the innovation enterprise becomes in tackling major diseases the higher will be the call for funding for its products. Technology—drugs, diagnostics, devices, and digital tools—will inevitably be in competition with other calls on the health system, such as medical staff and facilities. It will always be more politically expedient to cut the former than the latter.

Therefore, when engaging with policy-makers, the innovator community needs to take a much more holistic approach. They need to ensure good horizon-scanning, so that new technologies do not catch budget holders unawares. They need to analyse and argue for the changes in care pathways enabled by the technologies in the pipeline. They need to demonstrate how outcomes will be transformed, affordably, including how new technology can be used in a more targeted and responsible manner.

In short, the life science industry must help politicians and health systems to develop a precision medicine outlook, and establish the economic case for it. Otherwise we will simply see more of the kind of 'trench warfare' that is common in the European markets, and growing in the USA, where payers take a tough stance on all expensive new technology, because they feel they have no alternative if they are to meet their budgets.

Public–private research partnerships

One way in which governments can support a more effective and efficient research enterprise is via public–private partnerships, national and international. They can play a role in breaking down the innovation silos and challenging confrontational attitudes. They can conduct pre- or non-competitive research but also open up channels of communication on the challenges and directions of change in the innovation enterprise in which both are involved.

We see promising moves on both sides of the Atlantic. A leading example is the European IMI. It identifies areas of mutual interest, develops calls for academic proposals from across the EU and sponsors them, with project budgets typically in the region of several million euros. The companies making up the European Federation of Pharmaceutical Industry and Associations (EFPIA) also devote their own in-kind resources to pursue and guide the projects. (See Box 6.1.)

The USA has no equivalent of IMI (something I've heard policy-makers in the USA regret). However there are a number of public–private partnerships of importance for reforming the innovation process: NEWDIGS (the MIT-based pioneer of adaptive licensing); the Personalized Medicine Consortium; the TransceleratePharma consortium (working on shared methodologies to improve the effectiveness of clinical research); the Critical Path Institute (taking forward aspects of the FDA's Critical Path Initiative, especially in areas such as biomarkers); and CTTX, an effort to simplify clinical trials, based at Duke University.

There is therefore no shortage of public–private partnership-type effort on the agenda. If anything, there is now a need to converge the results of these various efforts on a few models that are accepted globally. Since medical innovation is a global

Box 6.1 IMI—the world's leading public–private partnership in life sciences

IMI was created in 2008 as a partnership between the EU and the pharmaceutical industry, in the form of EFPIA. The EU was conscious that the region was falling behind the USA in pharmaceutical innovation, and was keen to invest substantial funding in the development of pre-competitive technologies of relevance to the industry. The involvement of EFPIA in IMI's governance guaranteed industrial relevance and encouraged subsequent industry utilization of results. IMI had an initial budget of 2bn euros and has 50 projects now underway.

These tackle problems of shared interest in a wide range of diseases: neurological conditions (Alzheimer's disease, schizophrenia, depression, chronic pain, and autism), diabetes, lung disease, oncology, inflammation, tuberculosis, and obesity. One major project, colourfully titled 'Good Drugs for Bad Bugs', is addressing antimicrobial resistance.

It is now launching IMI 2, budgeted at 3.3bn euros (both budget figures include industry in-kind funding). An additional benefit is that IMI projects require collaboration between different academic centres across Europe. While this sometimes leads to bureaucratic complexity, it is certainly building academic bridges between universities that otherwise would mainly compete.

Many IMI projects are of direct relevance to implementing the seven steps described in the Chapter 5. For example:

- The molecular taxonomy of disease (PRECISEADS for inflammatory diseases and AETIONOMY for neurodegenerative diseases)
- The ability to combine electronic medical records from across the EU for pursuit of clinical research ('EH4CR'—the reader can guess the acronym!)
- GetReal, a project to explore the use of RWD in clinical development
- ADAPT-SMART, tackling the major implementation issues associated with adaptive pathways.

Future IMI 2 projects will add to this list. For example, one will develop methodologies for more structured patient input into benefit/risk assessment; another set of projects will develop outcomes databases to share across nations and companies. Such projects help build the intellectual infrastructure for the future of biomedical innovation.

business, we need to harmonize development approaches, regulations, and evidence needs as much as possible.

Life science clusters

The vast majority of productive biomedical innovation occurs in a relatively few geographical clusters. Boston, the Bay Area, and San Diego lead in the USA, but there are important smaller clusters in North Carolina, Florida, New York, and Chicago, for example. In Europe, the UK's 'golden triangle (London, Oxford, and Cambridge); Paris and Lyon in France; Heidelberg, Berlin, and Munich in Germany come first to mind.

Most of these clusters centre on major academic medical centres, but they also include a vigorous community of SMEs, and attract laboratories from major international companies. In fact, there has been a clear trend to relocate the internal R&D labs of these majors away from inexpensive but largely remote sites (like Pfizer's Grotton site in New England and its Sandwich site in the UK; and AstraZeneca's Macclesfield site in the UK) to the leading bioscience clusters (e.g. both Cambridge USA and UK).

The intensive local environment of clusters is of vital importance for sustaining innovation. They weave together the 'triple helix' of medical centres, entrepreneurs, and venture capitalists. Informal contacts between individuals give birth to new businesses, or forge new relationships. Most importantly, they allow people to advance their careers, moving between companies without moving house.

The phenomenon has been widely studied, to identify the key ingredients, including relevant centres (academic and medical), local talent, entrepreneurial leadership (the 'serial entrepreneurs'), and infrastructure. Scale is not the only criterion: dynamism is of equal importance.

'Clusters', of course, come at several levels. The most local level is what I call 'café clusters': a place where everyone in the area can walk to the same spot, and bump naturally into each other. A café in One Kendall Square in Cambridge (USA) hosts many such biotech conversations. In the early days of Genzyme, located just by Kendall Square, it is said that one café conversation involved an executive, who started to confide to a colleague a facet of the upcoming initial public offering (IPO) for the company. Suddenly, everyone else in the café fell silent!

Then there are 'cycle clusters', in which you can jump on your bike and quickly attend a seminar or hold an impromptu meeting with a colleague. Both Cambridges (USA and UK) would be examples of this. Widening, there are 'car clusters', where you can arrange meetings needing just a 30–60 minute car journey: the whole Boston/ Cambridge cluster within the outer ring road I-495 would qualify. Beyond that, one can speak of a regional cluster—for example, the whole North East USA cluster, but then it takes a full day to get to and from a meeting, or 2-day conferences to bring everyone together.

We can easily see that what can be achieved in each level is quite different. Also, the 'flavour' of each cluster reflects its history, culture, and the other sciences or industries in the area. The California life science clusters are in a position to bring in powerful IT companies at just the point needed for the information explosion of precision medicine. So there is no fixed blueprint.

The question policy-makers ask is: what can we do to create or grow a large cluster that is not yet at critical mass? This is not an easy question to answer, as clusters essentially grow organically, each new entrepreneur, investor, or company being attracted by the presence of others. However, it is hard for a cluster to form without some sense of identity, and so maps and marketing materials, laying out what is already there, have a role to play. So too can creating a natural geographical focus if one does not exist. The London cluster is impressive in total numbers of academics, companies, and investors, but there is no natural convening point. So MedCity was formed and is planning to use sites like the new Francis Crick Institute as a place for key figures in the cluster to convene.

The media

The media shape public opinion and political attitudes. But, with the exception of a few well-informed medical correspondents, the print and broadcast media tend to perpetuate a few stereotypes and narratives about medical innovation. These make it easier for the audience to quickly relate to their stories.

Some of these are helpful to the innovation cause, as when they highlight clinical researchers investigating exciting new developments that could bring breakthrough therapies to much-needed diseases. However, even then, journalists fail to distinguish early stage, speculative research (with a low probability of ultimate success) from those with a real chance of making a difference in lives of their readers. And of course sometimes attention-seeking scientists or university communications offices connive in this.

When it comes to commercial research, there is a well-rooted narrative that depicts companies, especially international pharmaceutical companies, as greedy, secretive, overly politically influential, and fundamentally untrustworthy. While of course, as in all human activities, all of these may apply to some people in some companies some of the time, this narrative is unhelpful to sustaining support for innovation. It makes academic researchers less willing to collaborate with industry. Those leaving for industry positions are often accused of 'joining the dark side'! However, such moves are an important part of keeping companies scientifically current. It also complicates issues such as the use of patient-derived data for commercial research, since journalists depict this as companies profiting from patient data, as opposed to helping them target future therapies to those who will benefit the most.

Without trying to curb legitimate interest when things go wrong—results hidden from scrutiny, unscrupulous marketing ploys, and so on—we need a narrative that positions human disease, not the companies trying to address it, as the enemy. And media can promote a greater genuine understanding of the major challenges these diseases pose, and the investments—both public and private—that we need to defeat them.

Media attitudes towards regulators is also very influential. They are consistently positioned as 'watchdogs', as we have seen earlier. The fact that some are funded by levies on industry is depicted as putting them in the pockets of those they are meant to regulate. As anyone close to this relationship knows, this does not give any specific advantage to innovators seeking to gain approval—in fact it tends to drive a very sanitized stance, in which senior figures on boards of bodies like the FDA, EMA, and NICE must not have personal or financial links with companies. While understandable, this denies these organizations insights from current or past practitioners. Wherever possible, transparency about potential conflicts of interest is much better than trying to create a world in which no such potential conflicts exist.

There is also a deep suspicion often fostered by the media on the desirability of anything linked to genetics. This, of course, surfaced dramatically in Europe when genetically modified foods were first introduced. A Europe-wide ban resulted, on crops that are improving productivity and serving consumers in the Americas and the developing world. So far we have managed to avoid such a backlash against such concepts as gene therapy, which can of course be regarded as restoring normal gene function in patients lacking it. But the next step is gene editing, via the CRISPR technology, for example. Such approaches are already rightly being debated in the scientific community, especially when they are applied to human germ-line genetics.

This is a good example of where the scientific community needs to engage with ethicists in advance, rather than after something goes badly wrong.

Ethics

Ethics have always been inseparable from medical innovation, because of the stakes involved, often of life-and-death gravity. There are issues on which major religions, or society as a whole, take very firm views: human cloning is an example. However, many issues involve complex balances of benefit and risk, and these can only be addressed by calm discussion that involves not just scientific and ethical experts but also the people directly affected.

The principle of equipoise has underpinned much thinking on clinical trials. The concept is that, at the start there is genuinely equal likelihood of a positive benefit/risk outcome from both arms of a clinical trial, whether the 'control arm' be a placebo or—in most cases—the current standard of care.

The challenge, as we saw in metastatic melanoma, comes when there is no good treatment available for a serious disease and any new treatment offers some real hope. Is it then right to deny half the patients in the trial a chance of survival, in the interests of those who might benefit in the future?

Of course, if the situation is clearly explained to the patient and they elect to participate without knowledge of which arm they will fall in, then it is perhaps defensible. But some argue that, in the case of a disease that is invariably fatal, we already know enough about its course that a control arm is redundant.

The ethical issues associated with adaptive development are especially tricky. We are presenting to patients the option of taking a treatment about which there is more uncertainty than past 'approvals'. And this is against the backdrop that such past approvals are presented as 'safe' (despite the long list of side-effects listed in the patient leaflets that few read). It is therefore likely that they will see taking products with this greater level of uncertainty as a 'trial' rather than as routine treatment.

Human attitudes to risk are notoriously perverse, as our psychology gives more weight to dramatic outcomes that are very rare than to ones that are more common but less catastrophic. Public attitudes to uncertainty have been much less researched: we need to understand how people react to situations in which both benefit *and* risk are more uncertain.

In addition to greater ethical consideration to benefit/risk and uncertainty, ethical issues raise their head in several other areas of life science. Public attitudes to stem cell research and gene therapy are a case in point. A great deal depends on how the questions are presented, and by whom, especially when emotive terms such as 'embryonic' and 'genetic' are involved.

There is no substitute for clinicians explaining the science and its implications directly to the media and general public. Leaving this dialogue to politicians, regulators, industry, or activists can often be disastrous.

And the issues emerging from human genomic science are complex, and so ethical discussions must be well informed. It was tough enough—though relatively straightforward—to convey the meaning of the gene sequence as a 'recipe for you' when we thought that the sequence of AGCT bases was just that. We now realize that gene copy number, insertions, deletions, and rearrangements carry much of the disease-related information. We are also discovering epigenetics as the means by which we not only shape the expression of our own genome but also of successive generations—and perhaps even of those we interact with by our own behaviour! Rather than a prescriptive determination of who we are, our genome is a library of possibilities, or—if you prefer—our personal management challenge. Explaining all this in layman's terms will not be easy.

A particularly thorny ethical issue, with very practical consequences, arises from the new streams of personal health data: not just genomics but also medical record data and data derived from social media. Who should be regarded as 'owning', or at least controlling, health data? Who should be allowed to access this data and for what purposes?

There are concerns with both identifiable and de-identified data. Personally identifiable data should obviously be restricted to medical professionals taking care of the individual, but should the flow to them be controlled by the patient themselves? What about personal data stored in genomic databases? That is much more valuable if linked to personal health data. Again, who should control the matching?

De-identified data has enormous potential for research. Indeed many of the opportunities described in this book would be impossible without its use. However, concerns abound about privacy and commercial exploitation of private data. The majority of citizens, if asked directly, favour the use of their data for research, but the media can easily whip up a firestorm of concern on the issue, as they did in the UK recently, in the so-called 'care.data' controversy. (Government announced they would be using de-identified patient data in research and the media focused on the scandal of 'selling your data to pharmaceutical companies'.) Health data is probably the most sensitive of the data privacy and data use issues that concern us in the internet age.

There will be a particular set of concerns about genomic data. This is partly rational: the data will contain messages or clues to diseases or ancestry. We will not want insurance companies, employers, hackers—or even in some cases family members—to be able to interrogate it at will. There will also be a semi-rational concern about its confidentiality and security: it is 'uniquely us', as close to a biological statement of identity as we can get.

Therefore, while consumer genomics companies rush to invest in more sequencing and deeper interpretation, I believe that secure, trusted guardianship of the data will emerge as a primary concern and potential bottleneck, at least for some of the population. In meeting this need for that section of the population, we will need to satisfy a number of requirements:

◆ An organization they trust, composed of individuals of unimpeachable integrity

◆ An absence of profit motive, ruling out business models that sell on data

◆ Nationally located resources, recognizing that trust often stops at national borders

◆ Control by the individual themselves of any access to the data

Whether or not such 'Fort Knox' repositories of genomic data are created, we will need national and international protocols on access and use of health data. I have a strong point of view on the subject. I do not believe this effort should be led by governments, or even research bodies: it needs to be led by patient groups. Only they have, and can be seen to have, primarily the patients' interests at heart. The need is becoming urgent: if this issue is not tackled effectively, we only need a few additional horror stories of wrongful access or use of data, followed by a media storm, followed by a political backlash, and we will set back by years the use of health data in research.

This leads us directly to the question of how we should involve the public more effectively in the whole innovation process.

Public and patient involvement

There is now an increasing realization that we need to take 'public and patient engagement' from a fashionable 'feel-good' factor to become a fundamental part of our innovation strategy. As we outlined in the Chapter 5, patient engagement should become a mainstream part of decision-making—from research priorities, through establishing acceptable benefit/risk to deciding how best to ensure patient adherence to therapy.

But public and patient engagement has a broader role in establishing the policy environment. For too long public and patient involvement has been a nominal 'add-on' to a process dominated by the professionals and their concerns and drives. Patients have sat as observers—or at best advisors—in groups run by innovators, regulators, or reimbursement agencies. They have been invited, often as an afterthought, to speak at both industry and public sector meetings, usually last on the agenda, when many have departed.

This is not just unjust, it is wrong-headed at a very fundamental level. While patients and the public are under-informed and under-involved, there is little that politicians are likely to hear from them except their unmet health needs and their frustration when the innovation process fails to deliver.

Many of the changes we discuss in this book require political championship or support: to whom do politicians listen, most carefully and consistently? Voters. Patients and the public. The expressed needs of their individual constituents. The views of the patient groups that they are often asked to chair, and, of course, the views of the electorate in general. Unlike some other lobby groups, they are free from the suspicion of commercial bias and professional vested interest. They, of course, have their own biases—towards effective, affordable treatment—but that is one bias that politicians will find it hard to dismiss.

So NICE has its Citizen's Council, advising on how its policies can best reflect patient needs and views. The FDA now routinely holds 'patient-focused drug development' sessions in which FDA experts actively seek patient input to regulatory issues. They also use webinars, interactive chat room sessions, and a dedicated newsletter to reach out to patients. The EMA has created a dedicated Patients (and Healthcare Professionals) department. In a 2015 review of the innovation process and policies in the UK, there is both a strong patient voice at all meetings and also an open website for members of the public to contribute solutions.

On a number of the issues raised on this book—and health data security is a prime example—we need to create new channels for public dialogue. The alternative is a media-led approach, where politicians listen to the mass media when they pick up a story and then respond, often in a knee-jerk manner, to what they see as a public concern. That killed genetically engineered crops in Europe; it could as easily set back any of the developments we will describe in Chapter 7.

An agenda for regulatory science

Over the last decade there has been increasing talk of 'regulatory science'. The FDA, especially under Commissioner Hamburg, called for its urgent global development. They commissioned a study by the Institute of Medicine[2]. As a result a number of academic centres were set up at major US universities. The FDA also urged a global cooperation in that direction, and this call is being taken up by the MHRA and others in Europe.

Of course, regulatory science is a cluster of disciplines, not a single science. It involves clinical, physical, and social sciences, as it seeks to tackle questions of medical, technical, and ethical significance. It is an important adjunct to the changes recommended in this book, as it provides the theoretical underpinning for them. Otherwise, they become a matter of opinion or speculation. Examples of questions regulatory science should tackle are:

- How can RCT and RWD be best combined in the development of products?
- How to deal with multiple morbidity when approving treatments for a single condition?
- Can new–new drug combinations, or combinations of drugs and devices, be approved through a single process?
- What is the best way of factoring in patient judgments into regulatory processes?
- How best can adaptive pathways safeguard legitimate ethical concerns?

There will be no shortage of such questions: the shortage will be of 'regulatory scientists' to deal with them. As a young and still-diffuse discipline, there is not a strong cohort of scientists who see themselves within it. Hence any initiative in this area must contain a strong educational component, for example Masters' programmes and joint industry–regulator fellowships.

One promising facet, however, is that this discipline could emerge as an international effort of scientists committed to converging regulatory structures and guidelines. Lack of harmonization remains a major source of delay and redundant effort with no value for patients.

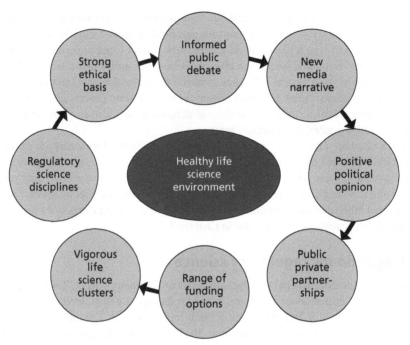

Fig. 6.1 Life sciences environment

The different facets of the environment we need are obviously not independent of one another. Without a strong regulatory science and ethical basis there cannot be a properly informed and positive public debate. Without this debate, it will be hard to change the media narrative, which inevitably influences many modern politicians. So the environment needs to couple action on all the fronts (Figure 6.1).

The need for a new 'social contract'

With so many issues of political, ethical, and media importance swirling around life science innovation, my CASMI colleagues have argued that we need a fundamental rethink of the 'contract' between society and the medical innovation process[3].

Of course, the current 'contract' cannot be found written down anywhere. It has no legal force or form: it is a general understanding across society of what the innovation enterprise should achieve and what are desirable, acceptable, and unacceptable way of achieving it. It is also not a contract between two parties: there are multiple stakeholders in the process. Each has an understanding of their role and responsibilities, and those of the others (or they should).

My personal perspective on key elements of the historical 'contract' are shown in Box 6.2.

'Society's' role in this contract is rather unclear: it is the context and milieu in which the debate takes place and the new understanding emerges. Principles, standards, and specific trade-offs are typically debated and decided by politicians, the agencies that report to them, and professional societies. So innovation policy is part of

Box 6.2 A personal perspective on key elements of the historical 'contract'

'Medical innovation is highly desirable, as long as it significantly advances patient benefit in an affordable manner. The various players in the life science enterprise should conduct themselves according to these principles and presumptions:

Life science research is worthy of public support beyond the level that academic curiosity would otherwise suggest, because of the long-term benefits that will flow to society. While this long-term impact is the goal, only leading researchers are in a position to judge what research to fund—via the peer review process—and the results should be similarly judged, by scientific peers, via independent journals.

Clinical trials are necessary to yield practical innovations but should proceed with considerable caution, for the protection of the vulnerable patients concerned. Research on future treatments should not take precedence over current treatment demands, which are the primary goals of funding the health system.

Companies are allowed to patent, and to make profits from innovations, as long as the products represent major advance on those currently available and profits are not excessive. They are expected to tell the truth about their products and if the products are found to cause significant problems companies should submit to legal challenge and penalties, whether or not they were aware of these problems when the product came to market.

Regulators are there primarily to ensure that unsafe treatments do not reach patients. They must guard against commercial companies introducing inadequately tested or flawed products. In determining acceptable benefit/risk, clinical opinion should outweigh any patient demands, since only clinicians can be sufficiently dispassionate and objective. It is more important to be sure than to be quick when it comes to approving products.

Companies need to justify their prices on the basis of the effectiveness of their products alongside other current calls on healthcare expenditure. At the margin, greater benefit of the doubt should be given to promising innovations that affect those suffering from life-threatening conditions.

Health systems and patients, while the ultimate customers for innovation, are not expected to play an active role in determining which innovations are needed and what benefits they should bring. This should be left to the researchers, both public and private. Except for some public health priorities, like vaccination, it is not the role of the health system to drive the implementation of innovations.

Data about treatments and outcomes are important but are the property of doctors and hospitals who are expected to keep them confidential.'

the political process, reflecting the ethical issues and public attitudes that we have described. However, the reality is that the politicians concerned are rarely skilled and experienced in life science innovation (there are some notable exceptions, of course). They are therefore dependent on the 'experts'.

While there are obviously aspects of the current implicit 'contract' that are unarguable, others need to be challenged. For example:

- Should strategic research goals in life sciences take more explicit account of population health needs, at the expense of scientific curiosity?
- What is the role of clinical research in a busy, economically pressured health service? Is it a vital activity, to be protected? How can important research receive enough priority, for the sake of the future patient, when budgets are squeezed and doctors are busy?
- Do regulators have a more proactive role to play in advancing innovations that address unmet need?
- Should extra support and more generous pricing be allowed for those innovations that seem to open up new avenues for future research and treatment, or just be judged on their current contribution?
- Should any account be taken of the potential long-term economic contributions of products once their patents have expired and prices dropped to generic levels?
- Should life science companies be seen as partners in innovation with the health service, or suppliers of expensive commodities?
- Should reductions in development time and costs result in lower prices?
- Should patients and patient organizations have a more active and influential role in determining and advancing the innovation agenda and in deciding on the acceptable benefit/risk ratio of new products?
- Is avoidance of risk the pre-eminent consideration when deciding what trials should be done and what products approved for use?
- Should health data be the property of patients, for them to determine access and use?
- How should we set limits on ethically acceptable research, in an age of genomics and gene therapy?

If my colleagues are right that the social contract for medical innovation is overdue for overhaul, these are the kinds of questions we should now debate. For the changes described in the Chapter 5 to flourish, we need greater understanding and trust between participants and the development of a common agenda.

Funding

Commercial bioscience funding is like a rollercoaster. For a few years investors are cautious, IPOs dry up and innovators struggle to get early stage companies off the ground. Then a few of the larger biotech companies thrive, a larger number of new products are approved, investor confidence improves, there is a steady flow of IPOs—but innovators still struggle to get early stage companies off the ground!

As I write, we are in one of the most sustained positive periods. The Biotech Index has outperformed the S&P 500 over the last 2 years, with only the occasional fluctuation. Over this period there have been more than 120 IPOs in the sector. The annual

JP Morgan Healthcare conference, which provides a bell-wether of industry sentiment, has been upbeat for the last 3-4 years. A recent European industry survey showed unprecedented levels of confidence—of continuing IPOs, continuing hiring, continuing lucrative mergers and acquisitions activity.

However, early stage innovators—particularly in Europe—still complain of the challenge in getting their companies funded. To understand this, we need to look at the sources of funding, where these are currently going, and what might be the particular problems of companies seeking to break through to financial self-sufficiency.

In fact, there are two funding challenges innovative companies face. *The first is the funding to reach proof of concept*—initial trials that demonstrate the effectiveness of the product. There are venture capital funds for this step—many more in the USA than Europe—but typically oversubscribed, and many innovators resent the loss of equity that typically occurs with venture capital funding.

Targeted 'non-dilutive' funds to reach proof of concept are very valuable for innovators. The UK's Biomedical Catalyst fund (jointly set up between the Medical Research Council and InnovateUK) has proved to be both popular and successful. However, more efforts along these lines would help the sector forward, and the recently launched Oxford Innovation Fund (at an impressive £300m) is a good example of the kinds of additional funding vehicles we need. The Swiss market, funnelling money from other countries, is also proving vibrant.

The second tough development stage is to reach commercial scale as an independent medium-sized company. This continues to be Europe's most serious problem in building an industry to rival the USA. Large-scale clinical trials and a global commercial infrastructure cost tens or hundreds of millions to build, and funds on this scale are very hard to raise. The temptation to sell out once a promising product is near commercialization, typically to a US-based major, is often irresistible to investors.

A good example of this phenomenon is Solexa, the Cambridge (UK) company acquired in 2006 for $600 million by Illumina. The Solexa technology has enabled the latter to catapult to become a company selling more than $2 billion in gene sequencing equipment annually. Illumina, recently voted the world's most admired company, is now worth more than $30 billion in market capitalization.

To secure the funds to take European start-ups to sustainable status, we need to tap into the huge financial resources in European investment markets (notably London), into the sovereign wealth funds and other sources with extremely deep pockets.

It will also be interesting to see what the world's best known nouveau riche companies—Google, Amazon, Apple, etc—do in life sciences. Their capital base is truly astonishing, and their founders and executives have started to show an interest in healthcare and make forays into areas that fuse life sciences and IT, like Google's Calico.

These Silicon Valley giants have the ability to make truly transformative investments in life sciences: it will be interesting to see whether they do.

The environment must be right for the effective translation of science into both global scale businesses and high patient impact. This sets a demanding policy and public dialogue agenda for those countries that wish to flourish in life sciences. They must:

- Unify the policy agendas of health and innovation, such that the highest value innovations are taken forward as rapidly as possible and are then reimbursed and adopted by health systems.
- Provide non-dilutive early development finance to bring promising products to the proof of concept stage.
- Develop creative structures to attract major new finance into the sector, to enable the emergence of global scale companies.
- Develop and use public–private partnerships to create enabling technologies at a pre-competitive stage.
- Encourage the emergence and development of life science clusters, bringing together researchers, companies, and sources of venture finance.
- Set up public dialogues to set out technology possibilities, debate implications, and surface concerns.
- Develop an 'ethical infrastructure' for life sciences that tackles the thorniest issues, including ownership, access, and utilization of health data.
- Create a discipline of regulatory science, to tackle the major scientific issues raised by the changes needed to the innovation process, and to ensure as much international harmonization as possible.
- Bring patient groups into the centre of policy-making and of media interaction to change the nature of the narrative.
- Foster discussion of the 'social contract' governing medical innovation, tackling the historical issues, such as lack of trust, and anticipating new issues arising from new technology and innovation process change.

This represents a meaty agenda. Without it, the reforms we need to the translation of medical innovation will not prosper. Governments that are most receptive to this agenda will stand to benefit the most from the changes underway.

Summary of Chapter 6

Change needs the right environment to occur. Politicians, the media, ethicists, public and patient debate, funding agencies, and investors can all play a major role in supporting the future of medical innovation. Public–private partnerships and life science clusters are vehicles in which the new precision medicine can flourish. However, we need a new 'social contract' to underpin the innovation enterprise.

In Chapter 7, we look at some of the most important new technologies and treatments we will see if we can get the environment right and make the necessary changes to transform the practice of innovation. The impact we can make on life science translation could be transformative.

References

1. **Mazzucato M.** 2013. *The Entrepreneurial State: Debunking Public vs. Private Sector Myths.* London: Anthem Press

2. Institute of Medicine (US) Forum on Drug Discovery, Development, and Translation. 2011. *Building a National Framework for the Establishment of Regulatory Science for Drug Development.* Washington, DC: The National Academic Press.

3. **Horne R, Bell JI, Montgomery JR, Ravn MO, Tooke JE.** 2015. A new social contract for medical innovation. *Lancet*, 385:1153–4.

Chapter 7

Achieving future impact

How can we use our toolkit for redesigning the innovation process to transform the effectiveness with which we exploit some of the advanced technologies in the pipeline? What impact could precision medicine make on the great healthcare challenges ahead, via advancing the speed and reducing the cost of future treatments? Let us look ahead to the enabling technologies in the pipeline, and the diseases they will help us to tackle over the next 20 years, and try to answer these questions.

There is a huge range of new enabling technologies that are coming available that all call for—or enable—new approaches to the innovation process:

◆ Genomic medicine
◆ 'Multi-omic' analysis
◆ Combination pharmaceuticals
◆ Stem cell therapies
◆ Gene therapies and gene editing
◆ High resolution imaging
◆ Intelligent implantable devices
◆ Digital tools for personal health management
◆ 'Big data'
◆ Artificial intelligence and machine learning
◆ Clinical decision support

We have already introduced many of these as tools for change. This chapter explores in more detail how all these technologies could reinforce this new era of precision medicine. We then look specifically at a few areas of great human need that are poised for innovation-led transformation—cancer, rare hereditary diseases, the avoidance of chronic disease, and the diseases of aging, especially dementia. As we do so, we will seek to apply our seven 'steps to sustainability'.

Finally, we project the kind of overall impact we could make from transforming the process—impact that I estimate could increase 'net patient benefit' from bioscience by more than an order of magnitude.

Enabling technologies in the pipeline

Advances in 'omic' analysis, new advanced therapies, and information tools to guide them hold real promise in the design and delivery of treatments, many of which will have curative potential.

Genomic medicine

Over the last decade, we have accumulated a handful of genomic markers for guiding therapy, particularly in cancer. Gene panels, exome sequencing, and next-generation WGS are rapidly overtaking these, as their costs and time fall dramatically. And their use is extending from cancer to rare diseases and will increasingly be used to sub-segment the more common ones.

Everyone knows that the gene sequence is a binary code of life, present in every cell and passed on from one generation to another. So we tend to think that sequencing and interpreting the genome is a precise science. Not so, at least not yet. Different sequencing technologies still yield subtly but sometimes importantly different answers. And the interpretation of the variants that emerge is far from settled. An analogy might be reading a ticker tape scrolling across a stock price screen. By the time everyone has seen the result, it has changed, and the decision required may have changed with it.

The UK has taken a major step towards genomic medicine in its 100 000 genomes project. Over a 3-year period, it will conduct Illumina WGS on samples from around 60 000 patients and relatives. Cancer and inherited diseases will be the major focus, in an attempt to link the genomic nature of the disease and the effectiveness of the choice of treatment. Comparisons with the data in patients' medical records will enrich the analysis, which should reveal how specific mutations in tumour cells affect the development of the disease and the likely effect of chemotherapy regimens, and how rare diseases arise from inherited genetic differences. There will also be learning from the process itself, from how patients are consented and samples are handled and analysed to the way in which clinical interpretation is made and acted upon.

Over a similar timescale, 100 000 of the citizens of Saudi Arabia will also be geno-typed. Here there is a particularly sharp focus on inherited disease because of the very high level of consanguinity in the population: many marriages between cousins, as a result of the structure of Saudi society. This leads to an 8% level of inherited disease in Saudi newborns. The 'Saudi Mendeliome Project' uses ThermoFisher gene panels that cover 3000 genes associated with inherited disease. Initial findings suggest that the disease can be identified in around 43% of patients.

Whole exome sequencing (WES) was shown to detect a few additional patients but at a much higher cost ($2500) and longer time (3 months), compared with the $150 and few days required for the panel approach. This illustrates the cost and time tradeoffs that exist for any specific application between WGS, WES, and selective gene panels. It may be that panels can be used first on all suspected of inherited disease, those still undiagnosed studied with WES, and the remainder WGS. Or, as the cost of WGS falls further, or sequencing newborns' DNA becomes common practice, this may eventually become the first, not last resort.

The Saudi project is intended, like the UK's, to create the foundation for genomically based precision medicine in the country via a network of satellite labs throughout the country. The findings will have important findings for both science and for how their—and perhaps other—societies might evolve.

The genomic analysis of the Icelandic population has been a major initiative of recent years, led by a company called DeCode Genetics (now part of Roche). They have discovered interesting hereditary facts in this geographically isolated gene pool, such as the preponderance of mitochondrial genes from Scottish women, presumably loaded on to longboats after Viking invasions centuries ago[1]. But their work on WGS of 2636 Icelanders (plus gene array results on 100 000 others) has also resulted in a number of novel health-related discoveries. The gene pool is relatively homogeneous and so there is less 'background noise' in gene association studies. New genes associated with diseases such as early-onset atrial fibrillation, liver disease, and Alzheimer's have been identified[2]. They can also identify individuals with specific gene 'knockouts' (the human equivalent of 'knockout mice' studied in the laboratory) and find the health consequences of these[3].

In the USA, there is a veritable Californian Gold Rush into genomics. This began with the 23andMe offering to consumers of hereditary and health information based on specific gene tests. With the new companies set up by Craig Venter (Human Longevity), Lee Hood (Arivale), and, most recently, Illumina (in partnership with private equity and the Mayo Clinic), literally hundreds of millions of dollars are being channelled into genomics-based efforts. Many of these are oriented to the consumer. It remains to be seen how rapidly this enthusiasm for consumer genomics spreads globally. We are also seeing what some would regard as more trivial applications of genomics. One of the Duran Duran pop group, Nick Rhodes, has teamed up with Chris Toumazou, Regius Professor of Engineering at Imperial College London, to create GeneU, a company that will run a gene chip test in 30 minutes and determine which two of 18 serums should be combined in a skin cream, exactly targeted to the genetics of your skin. As public awareness of genes and their importance grows, we will probably see many more such ventures[4].

Serious disease-focused efforts, however, already promise advance in areas in which we need more molecular level clues to mechanism. The company Biogen has teamed up with Columbia University Medical Center and the ALS Association (the patient organization) to study the genomes of 1500 ALS sufferers. They hope to understand the genomic basis of different clinical forms of the disease and so be able to better target clinical trials.

In a book in which we are exploring ways of being more flexible and streamlined in regulation, it may seem strange to suggest introducing a new area for regulation. But that area is genomic medicine. There are multiple places in the process of determining a gene sequence in which error or artifact can occur.

Patients must be properly consented for the results to be used. Samples must be taken, stored, and transported correctly. DNA extraction can go wrong. Running the samples on the sequencer may be almost foolproof, but deciding to what 'depth' it runs (the number of times any sequence is sampled) is not straightforward. This needs to be adjusted to match the application. Then there are a variety of algorithms for variant calling, which are being refined all the time. Finally, the clinical interpretation is far from trivial for many diseases. So the process has multiple decision points and potential sources of error or difference.

We have already seen some of this in the early days of the UK's 100 000 genomes project. One specific challenge is the choice between fresh/frozen tissue and paraffin block-embedded samples (the usual medium for pathologists). The first sounds more attractive from the layman's point of view—less chemical treatment, closer to the original tissue, etc. But, as one surgeon pointed out to me, the quality, handling, and rapid processing of the tissue depends greatly on whether the surgeon cares about the genomic analysis. Some do, some don't. The latter may leave tissue to degrade before it ever reaches the lab.

These potential failure points led me, with CASMI colleagues, to propose the concept of Good Genomic Practice (GGP), paralleling Good Laboratory (GLP), Good Clinical Practice (GCP), etc[5]. The logic is simple. If practitioners do not propose the right best practice framework, regulators—who are often forced to act in response to major problems that reach the public's eye—will set it. Such regulation could either inhibit the field or just focus on the wrong issues.

The FDA has already started consulting on regulations for WGS, but this effort focuses (not unexpectedly) only on the 'test' itself—the sequence of events from sample entry to sequence results. We need to determine and spread best practices both before this point—in sample selection and handling—and after—in variant calling and clinical interpretation.

We have the opportunity to build genomic medicine globally from its inception. This is the aspiration of the GA4GA Alliance, in which institutions and countries will be pooling genomic data for specific conditions: it will be vital that data protocols as well as disease definitions are consistently defined.

Multi-omics

The fast-developing field of personalized or stratified medicine is built not only on genomics, but also on a range of other tools. To recap, epigenomics, along with transcriptomics, studies the control of gene expression and the extent to which genes are expressed in the various tissues of the body—in health and in disease. Proteomics measures the level of the various proteins that result. Metabolomics defines the profiles of metabolites circulating in the blood, or—via functional MRI and other in vivo techniques—the levels present in specific types of cells. Techniques for immunophenotyping patients can quantitatively assess the state of the immune system, as it deals with infection and other disorders. And finally micro-biomics catalogues the type and numbers of micro-organisms that cohabit with us, to beneficial, harmful, or neutral effect.

Such approaches have been termed 'multi-omics'. Just as we saw in the work by (and on) Mike Snyder, we can use a variety of tools to accurately map and track many facets of the patient's biological status quo. The maturity of the technologies making up the field is, however, very variable. Although sequencing technologies for the genome are reasonably consolidated into a small number of platforms, software for sequence determination, and methods for variant calling and so the derivation of clinical interpretation are all far from standardized. Epigenomics is also far from a standard technology, although we can now measure the levels of methylation (one form of epigenetic control) on a gene-by gene basis.

Proteomic techniques to both measure protein levels and identify amino acid substitutions resulting from gene mutations are comparatively well established. We have been measuring metabolites in the blood for many years: these are the basis for most of the routine 'labs' ordered by clinicians to assist in diagnosis. However, 'metabolomics'—a more sophisticated and comprehensive profiling of all the metabolites—is a relatively new science. Immuno-phenotyping is also comparatively new, but gives us a profile of that most central of all human systems, the immune system, involved both in multiple diseases and also natural mechanisms for fighting disease. However, standard techniques and display of results are still some way off.

Microbiomics profiles the microorganisms that co-habit our bodies and whose significance is only now being discovered, as is its involvement in various diseases, not just infectious disease.

One term being given to the integrated use of the various 'omics' technologies: iPOP—integrated Personal Omics Profile. One can imagine a future in which not only do we have our genomes sequenced at birth but we also track our iPOP annually to ensure we see the early emergence of any disease and take preventative action, which is nearly always more effective (and cheaper) than late intervention when symptoms have appeared.

The fast development of multi-omics poses a major issue. With so many types of measurements being made on upwards of 10 000 different molecular entities, using so many types of equipment, how can we bring order to this field? What hope is there of a standardized approach that enables us to compare and contrast the molecular profiles of different individuals so as to inform the care of each? Unless we can do this, all we will have is a series of fascinating anecdotes, but no narrative to weave them together.

Fortunately we can tackle this problem with the aid of machine learning or artificial intelligence. Linkages between what we find in the genome and via other 'omics in particular patients—and particular disease states—can be analysed by the kinds of techniques used in linguistic translation or in mining massive databases. We will return to these possibilities later.

Combination therapies

Doctors have prescribed a number of drugs or other treatments together for a very long time, conscious of course of the risks of cross-reaction. Many clinical trials evaluate a new drug in combination with the standard treatment. However, as we learn more about diseases such as cancer and inflammatory conditions, we realize that a combination of two new products may be superior to anything in current use.

Mutations appear in tumours with complex 'fingerprints', and the best regimen of anti-cancer drugs for a particular patient, or group of patients, needs to align with those fingerprints. There is also an increasing trend to combine anti-cancer 'designer drugs' with immuno-therapies that alert the body's own immune system to the presence of the cancer cell.

Traditionally, regulators have been cautious about 'new–new' combinations, preferring each new drug to prove its worth on a standalone basis. This will need to change if they are to keep up with drug discovery and genomic targeting.

Stem cell therapies

These same regulators have faced the emergence, over the last few years, of cell therapies that could offer regeneration or bolstering of failing capabilities in the joints, brain, heart, or other organs. For the most part they have been flexible as the therapies enter clinical trial: there were 123 such trials underway when reviewed in 2011[6]. These are typically in either a single centre or a very few centres skilled in producing and handling the cells.

Of course, those stem cell therapies that prove their worth should be rolled out across the healthcare world. Therein lies the larger problem. Can the precise protocol used in centre A be replicated in B, C, etc? And, even if it can, are the cell preparations that result, in fact, identical. The answer to both questions may be 'no', because of the impact of very subtle differences in the starting cells, culture conditions, and handling. Cells are not inert chemical substances but living entities, whose interaction with each other and with the surfaces of vessels—let alone the characteristics of the patient into which they are injected—are capricious and unpredictable.

We have to distinguish autologous cells—those taken from the patient themselves and then manipulated in some way before reinjection—from 'standard' products from allogeneic sources such as cord blood. The first EMA-approved stem cell treatment, Holoclar for the repair of eye damage, is autologous: stem cells are removed from a healthy region of the cornea, grown in the lab, and then transplanted[7].

Perhaps the most exciting autologous therapies under development are the 'CAR-T' engineered T-cell preparations that companies such as Novartis and Juno have begun to develop. These modify the proteins on the surface of the T-cells, already tuned to attack foreign cells such as tumours, to ensure they recognize and engage with the specific cells identified in the patient.

Some of the development issues that arise in this field have begun to be tackled by consortia such as the CTSCC (CASMI Translational Stem Cell Consortium) and the ARM (Alliance for Regenerative Medicine). Here a variety of organizations and companies, both those innovating cell therapies and those providing equipment and reagents, see a common imperative to set standards, anticipate regulatory challenges, and to evolve the ecosystem productively through their partnerships.

Cell therapies will share with the gene therapies we are about to describe one huge challenge—creating viable reimbursement models. One-shot therapies that may cost (or be priced) at $100 000 or more per patient may seem unaffordable to health systems even if they replace many years of costly supportive care. New financial instruments, of the kind described earlier, may be needed here.

Gene therapies

Gene therapy has been a 'gleam in the eye' for therapeutic medicine since the late 1980s, when disease-specific genes began to be identified and the concept emerged of delivering them into deficient cells using viral vectors. Frankly, progress since then has been disappointing, as a result of some challenging technological problems and a death in a gene therapy trial that set back confidence in the field for a decade.

Gene therapy is a simple concept—replace a deficient gene with a functional one—but fiendishly difficult to implement. Delivering the gene to the nucleus of a target cell is the most challenging barrier. Viral vectors can achieve this, delivering 'bare' DNA to target human cells that lacking them. Two promising examples are:

◆ Early clinical trials for adenoleukodystrophy. This severe genetic disease that results in nerve demyelination results from a deficiency in a transporter protein coded by the *ABCD1* gene. French researchers have begun to implant stem cells that have been corrected by a lentivirus-delivered ABCD1 into 7-year-old boys with early signs of the disease. They demonstrated that the protein was being expressed in a proportion of blood cells and that disease progression stopped after 15 months[8].

◆ The delivery of a miniaturized form of the dystrophin gene to the muscle cells of boys with Duchenne muscular dystrophy. US researchers have used adeno-associated virus to carry the gene into muscle, and shown beneficial effects at both the molecular and symptom level[9].

Gene-modified T-cells are able to recognize and trigger the death of target tumour cells. We will describe its potential in cancer later in this chapter.

Gene editing technology is a much newer concept. The CRISPR technology (clustered regularly interspaced short plaindromic repeats) was first applied to human cells in 2012, but it has quickly sparked both research and commercial interest, with the formation of a number of companies, and the start of a likely long patent dispute between scientists at MIT and others at UC Berkeley.

The CRISPR approach will first be a research tool, using RNA-guided nucleases, derived from bacterial response to invading viruses. These have been shown to modify genes in a variety of cells, intervene in gene expression, and label specific loci in the genome[10]. The technique can be used to establish the link between specific genes and cellular functions, expanding the field of functional genomics. It is also being applied to the development of cellular and animal models containing specific genes, expanding our in vitro screening capacity.

To become useful as a therapy, of course, the edited gene must reach the cell type of interest. So, like other forms of gene therapy, the CRISPR editing 'kit' has to be delivered to the target cells via a viral vector like adeno-associated virus.

In the long term, CRISPR technology undoubtedly has quite broad potential, as laid out by one group of its founders[11]. This paper also describes the 25-year step-by-step journey from the first discovery of odd repeats in bacterial genomes to the development of tools for human therapy—a fascinating story, with clear lessons for the support of basic, curiosity-driven science. In principle, CRISPR could correct monogenic disorders such as cystic fibrosis, sickle-cell anemia, or Duschenne muscular dystrophy. It could also treat diseases that result from the duplication of gene sequences, such as Friedrich's ataxia. The elimination of the HIV genes incorporated in cells is a further target.

CRISPR is just part of a burgeoning field of 'genome engineering', in which gene insertion, deletion, and inactivation are achieved using a range of enzymatic tools, increasingly quickly and reliably. It is possible, as George Church has shown, to automatically create billions of genomic combinations in a multiplexed desktop instrument, effectively 'speeding up evolution' by selecting those with desired properties[12]. So far such 'genome engineering' tools have mainly been used in microorganisms, with the potential to increase the yield and modify the output of biopharmaceutical manufacturing[13].

All gene therapy treatments, like most cell therapies, will face the reimbursement challenge of a high price, one-shot treatment that avoids lengthy palliative treatment and often prevents tragic shortening of lives. But the costs, both in terms of the materials and the procedure to insert them, is likely to test constrained budgets. The creative reimbursement thinking, advocated in the Chapter 6, is clearly going to be needed.

Nanobiotechnology

The ability to manipulate biological materials at the nanometre scale has huge potential in research tools and diagnostics. It can be used in at least two senses: to apply techniques drawn from biology to create ultra-small scale 'machines'; alternatively, to apply ultra-small scale tools to the investigation and manipulation of biological systems. Like most 'hot' scientific fields, the definition expands to fit the funding available!

There are potential applications in both diagnostics and therapeutics. Attaching antibodies to semiconductors can increase sensitivity and specificity of antibody tests and modified magnetic nanoparticles can highlight tumours or other areas of interest in the course of surgery[14]. In therapeutics, drugs or gene therapies could be delivered by targeted nanomolecules.

High resolution imaging

Imaging has already had a major impact on medicine, of course: CT, MRI, and—to a lesser extent—PET scanning have become staple methodologies in the management of many diseases.

However, as sensitivity and sophistication increase so do the possibilities for using imaging as a very sensitive means of tracking the progress of a disease and its response to treatment.

One modality of rapidly increasing power is dynamic contrast MRI, in which a contrast agent, gadolinium, is injected into the patient and enables the measurement of blood perfusion in diseased areas, such as inflammatory lesions and tumours. These measurements, such as the DEMRIQ score, enables clinicians to assess quantitatively the way in which blood flow changes in the early stages of treatment response[15]. This can give readout on efficacy much earlier than more gross measures such as volume of tumours or inflammatory lesions. A pixel-by-pixel image, captured over time yields a more accurate answer than a radiologist's by-eye impression. Such tools need to gain regulatory acceptance as surrogate endpoints, since they provide rapid and accurate feedback on treatment effectiveness.

Functional MRI (fMRI) tracks the level of specific metabolites in tissues, and this is in the process of transforming the understanding of normal function and of diseases in the brain. It maps areas of metabolic activity via increased blood flow (although researchers are still arguing about the exact mechanism of the enhanced signal from active areas). As well as a tool for examining normal brain function, it can be used to assess damage from strokes, or the pattern of vascularization around brain tumours.

As therapies for brain diseases such Alzheimer's are developed, such tools will be vital to track their impact on brain structure and functional health[16].

Intelligent implantable devices

In a sense, these are already with us, in the form of implantable defibrillators, sensing the interruption of the heart rhythm and stepping in to restore it. The most sophisticated of insulin pumps for diabetics now sense and respond to blood glucose measurements.

However, we can easily foresee other devices that measure the levels of analytes in the blood, calculate their significance, and dispense one or more therapeutics, then check that the required changes have occurred. In psychiatric diseases, such as epilepsy, an implantable device could warn of seizures and be activated to correct them.

New classes of innovation always pose regulatory questions. Would these developments be devices, or pharmaceutical products, or some new category entirely, from the regulatory standpoint? The strength of the European regulatory system is its consistency across the member states of Europe. However the weakness is that discussion to amend or create legislation is painfully slow, so it may be best to utilize the device route to the maximum extent possible.

There is also an emerging a completely new regulatory issue associated with such devices—the threat of hacking. How can manufacturers guard against this, if a prominent individual is known to have such a device? If we (or rather they) can hack into cell phones, there would be a very real concern that a skilful hacker could disable a device to attack someone of whom they did not approve.

Overall, we will need to create a new route for the clinical development and regulatory examination of such complex and ambiguous products. However, the principles that we have already laid out should guide their shaping and use.

Clinical decision support

As the complexity of both diagnosis and therapy increases, we will face a huge challenge in guiding doctors' decision-making. Genomic and other omic data will help pin-point the patient's disease at a molecular level, but the amount of raw data and the complexity of interpreting it will be beyond the grasp of many—perhaps ultimately all—treating physicians. Once the disease is fully described, many of the therapies will be combinations of drugs that require monitoring protocols to ensure they are proving effective.

Of course, medical education (both initial and continuing) needs to grapple with this problem. There will be no widespread genomic medicine, for example, if there is no genomic medical education. But the speed of advance means that no formal educational tools can keep pace. This reinforces the need for intelligent clinical decision support (CDS).

To be effective it will need to complement, not replace, clinical judgment. Doctors will react sharply against a system that tells them what to do, and regulators will apply very strict standards to any system that claims to provide *the* answer. We need to present the data in a readily assimilable form and show how they lead to the treatment options that should be considered. Also, a single presentation of data and options may not do for all treating physicians. Highly research-aware doctors will welcome detail that their hard-pressed, time-constrained colleagues may not want. Effective CDS systems need to be customizable.

The last 20 years have focused heavily on the comprehensive electronic health record. Without the base data, mistakes will be made. But what doctors need is not a presentation of everything the electronic health record contains, but what is significant. Here machine learning can help refine the contents of the future CDS.

Clinicians, as my colleague Ian Abbs puts it, are pattern recognizers. With the patient in front of them, they typically do not work through complex decision trees, of the kind seen in clinical guidelines, valuable as these are for capturing best practice in a field. The pattern a doctor sees will include the key facts and options served up by the CDS, as well as what they see and hear from the patient in front of them. If we are to practice precision medicine, as laid out in this book, the CDS will become increasingly important.

Advanced diagnostics

As described in Chapter 4, diagnostics has long been the Cinderella of the life sciences sector, underfunded, and of little interest to investors. This is now changing dramatically. With the application of miniaturization, often to the nano scale, with the precipitate drop of the cost of WGS, the ingenious range of arrays for detecting specific genes and their expression, and the more routine use of mass spectrometry and arrays in proteomics, there is no shortage of exciting new diagnostic concepts.

One was launched in 2003 by a Stanford dropout, Elizabeth Holmes, whose company has developed a test that determines 30 analytes (out of a possible total of 200) from a single drop of blood. This has made her the youngest female billionaire in the USA (which probably means the world). Her company, Theranos, became worth $9bn, even though most experts in the field don't know how the microfluidic technology actually works! It seems likely that it is the energetic retail concept—a lab in every high street drug store—was mainly responsible for investor enthusiasm[17].

In fact, particularly for WGS, the cost of 'running the test' is now small in relation to the cost of its interpretation. With the onrush of new technologies and their ability to deliver multiplexed results, this shift in the cost balance may soon be a general phenomenon.

Digital healthcare tools

The digital revolution that is transforming so many aspects of our lives is now poised to create innovation in healthcare and, in fact, to empower many aspects of precision medicine. Broadband connections, mobile smartphones, telemedicine, social media, miniaturized wearable health sensors, and portable diagnostic tools all have a part to play, along with the electronic medical record, of course.

The goals of digital health will also be many and varied. For patients, the support of self-care, monitoring them as they live their lives, at the home, the office, or the gym, and reminding and stimulating them to adhere to therapy. For the professional, as we have just seen, we need to deliver much more flexible and intelligent decision support tools. These should sharpen the clinician's understanding of his patient, present best practice guidelines in treatment choice, and provide monitoring information to ensure the treatment is going to plan and having the desired outcomes.

Remote monitoring of patients via wearable devices and small in-home monitors are now receiving extensive trials. For conditions such as COPD, congestive heart disease, and diabetes, such remote monitors can anticipate and avoid the crises that currently result in emergency room visits and hospitalization. One example is continuous glucose monitoring from Senseonics: a tiny sensor is implanted in the upper arm of a Type 1 diabetic and the measurement transmitted to the patient's smart phone or a medical facility. This is now in use in the USA and UK and replaces the diabetic's need to regularly puncture the skin and use a kit to measure glucose.

However, the potential to put miniaturized sensors to work in healthcare has attracted the attention of those using such techniques in other areas of life, such as Formula One racing. As the driver speeds around the track at over 185 mph, hundreds of sensors give real time feedback on the driver's pulse and blood pressure, as well as the automotive metrics of engine and tyre temperature and oil pressure. As many as 1 to 2 gigabytes of data are generated per race. The potential of such sensors to monitor surgeon movement and performance is already being investigated by Maclaren. The company is now partnering with GSK to leverage its racing telemetry in clinical trials, focused on the patient movements that can continuously monitor their response to treatment in diseases such as stroke, ALS, Parkinson's, and RA.

We will see many such sensors developed. While there may be regulatory issues, the greater barrier could be reimbursement, for which a sound evidence-based economic case will need to be made. On this dimension, the initial results have been mixed. One study found no significant outcome differences between two populations of heart failure patients, one of which received such monitors. It is also asked whether the health system can deal with the alerts that are created and resource the staff needed[18].

Monitors will also enable telemedicine: medicine at a distance. A tele-intensive care unit study, in which patients in several centres were monitored by intensive care specialists from a single centre, showed benefits in only the sickest patients[19]. So we need not only more work on the design and operation of remote monitoring devices, they need careful economic evaluation if they are to attract the support of payers.

Mobile devices

Since smartphones are increasingly the way we manage our lives, and we buy them ourselves, they may well ultimately have the greatest impact. Their high portability and multi-functionality creates a new, fast-evolving dimension to our healthcare systems. As the healthcare world and individual patients realize the power of these 'personal health assistants', they will move from curiosities to being imbedded in pathways of care and daily health maintenance. We already have about a hundred thousand health-related 'apps' for mobile devices, and there is no sign of app ingenuity flagging. For some examples of popular apps, see Box 7.1.

With the plethora of apps available, the English NHS has set out to 'certify' some of them. Criteria include relevance to the population, the use of trusted information, compliance with data protection, and clinical safety. Early examples of certified apps deal with mood, fall prevention, pregnancy, and other health system-oriented issues.

The ability to embed tiny accelerometers on smartphones or smart watches does not only open the continuous measurement of routine exercise measurement. More sophisticated 3D movement tracking helps diagnose children with mobility problems, such as those caused by Niemann–Pick's disease[20]. Instead of the burdensome and

Box 7.1 Health-related apps

Fitness seems to lead the pack. Although there are very large numbers of consumers wearing dedicated activity trackers like FitBit or Jawbone, mobile phone apps from MyFitnessPal, RunKeeper, Azumio, Nike+, Runtatstic, and Endomondo all have at least 16 million downloads.

Other leading apps in US media are mainly focused on fitness and wellness: Fitocracy (workout levels), Couch to 5K (running), Fooducate and CarbsControl(diet), FitStar Personal Trainer (personalized exercise plans), HealthyOut (healthy eating out), Calorie Counter (matching diet to exercise), ShopWell (healthier shopping), Waterlogged (liquid intake), and Nudge (integrating other apps into an overall wellness score).

There are multiple disease related apps, although the connection to prevention via lifestyle is strong in most of them[a].

[a] Dellinger AJ, 'Come on, get healthy: our favorite health apps for the warm weather ahead', Digital Trends (2015), http://www.digitaltrends.com/mobile/best-health-apps/, accessed 15 Jul. 2016.

episodic 'six minute walk test' to measure the patient's mobility, such tools give continuous feedback on the level of movement and nature of their gait.

While most mobile applications remain 'sub-clinical' in their use, not affecting what the health system recommends to the patient, this 'Wild West' can probably continue. But the real power of these digital tools will come when they recommend, trigger, or even dispense clinical actions, and this is when the regulators and payers will really need to take notice.

Social media are becoming the lingua franca of the upcoming generations. Monitoring of social media can identify impending infections and both the benefits and side-effects that patients experience with new treatments. Regulators are even considering analysing this source of 'low grade but high volume' feedback on the safety of medicines in routine use.

What are the translational barriers that could prevent all this digital potential from transforming our health lives as it has so many other dimensions of our existence? Two main ones that I see—firstly, the challenge of creating viable business models based on reimbursed products and services, and secondly, the failure to integrate the various tools and data sources around the individual.

In most areas of life in which digital technology has made a major impact on our lives, there have either been strong consumer pull and payment, or clear economic or user convenience triggers for large organizations serving those consumers. Most of the mobile phone revolution is consumer-funded, as is music and movie downloading. The impact in industries such as retail, travel, and other services comes about because companies can offer a level or immediacy of service with no increase—and often a sharp decrease—in their costs.

In principle, some digital tools can enter patients' lives in the same way, by purchasing relevant apps. The real investment comes in tying in these and other mobile tools, such as remote monitoring devices, into the care process—and adapting the thought and work patterns of clinicians to absorb the new technology.

In many cases, the benefits are in terms of improved outcomes rather than reduced costs, and the additional investment to 'go digital' initially comes on top of the costs of maintaining the current system. This will be a challenge for many state-funded health systems, even if over the longer term digital tools will enable much more cost-effective pathways of care and prevention programmes. Consumer choice with economic consequences drives most digital revolutions. If consumer choice between providers exists, this can work in healthcare, and going digital in one form or another becomes a cost of their staying in business. But when consumer choice has little leverage, progress may be much slower.

The second major barrier is fragmentation. Thousands of apps, scores of local electronic systems for booking appointments or accessing data, and fragmented electronic medical record systems in primary and secondary care currently represent 'islands of automation'. They hold the promise of an integrated information-led process of patient-centric care, but will they ever fulfil it? This danger has been seen since electronic records began to be in common use two decades ago, and in many parts of the world the problem has got worse not better. The failure of the major £15bn 'Connecting for Health' project in the UK's NHS is just one example of the difficulty of achieving this integration.

I believe that the key to solving this problem lies in the patients themselves. As they gain greater control of their own health data and also generate increasing amounts of data themselves, then they will increasingly demand and expect that we can connect two worlds: the world of professionally gathered data about them—principally their medical records—and the private data that each has generated. There needs to be a consumer-led revolution that wrests control of the data away from clinicians and their institutions and becomes increasingly intolerant of dis-integrated data about them.

Figure 7.1 illustrates the variety of data sources and databases that can hold value for an individual's health life and care pathways. For maximum impact they need to come together in at least three places—for the patient herself, for the medical practitioner who wishes to see all the relevant information for a particular treatment choice, and for the accountable care organization, which will want to monitor health outcomes and costs, to ensure it is getting value for money from the whole care pathway (Figure 7.1)

What I call the 'patient portal' will be key. It will need to be a highly customized presentation of a variety of metrics, from exercise and nutrition, contain prompts to ensure compliance, and enable the scheduling of appointments. It should also give access to educational modules that keeps the patient current on their disease (and/or their state of wellness) and the steps they and others need to take to manage it. Who will develop such a portal?—Apple, Samsung, or Microsoft? Or major health systems?

The portal will need access via a 'rules engine' to the various sources of individual data from care providers, diagnostic labs, genomic data stores, and apps in daily use. It will also reflect the way in which the individual thinks about their health, and contact with the groups with whom they like to compare their progress and share information. Gamification (one of the ugliest words yet invented in English) will be an important component. Teenage Type 1 diabetics could compete with each other on daily glucose

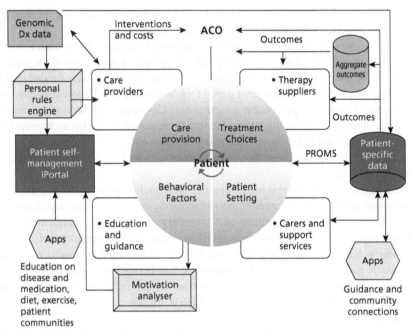

Fig. 7.1 The whole care pathway.

control; octagenarians can compete with their peers on exercise, diet, and medication adherence. It is often said that we are only able to maintain an activity that we know to be good for us for about 3 months. After that we become bored, and we need something to raise and sustain interest: many of us never lose our competitiveness!

Integration across applications is not just important for the individual patient (or healthy person). Healthcare providers need a different form of information integration, so that the signs and symptoms, diagnostic data, family history, and recent patient behaviours come together in a CDS system. Such a system can present choices (not decisions) to the clinicians that take these factors into account, perhaps with the aid of artificial intelligence that helps sort the wheat from the chaff.

For the system manager (the accountable care organization, 'ACO', in Figure 7.1) the individual detail will be less important as long as it is clear that the neediest and therefore most costly of the patients is receiving anticipatory, well-managed care, and that the overall system is delivering value for money, in terms of outcomes over costs.

Integrating health data of all types around the individual and delivering these in customized ways, will be a powerful force behind precision medicine, making it scalable and practicable. Many will see this picture as an unattainable goal, but unless we seek this ultimate integration, we will fail to realize all the much-vaunted potential of the digital age of healthcare.

As well as representing powerful tools in their own right, these digital developments can help bring about some of the changes needed in the innovation process for other types of medical intervention. This has value for several of our gaps in translation.

As we saw earlier, the past approach to monitoring clinical trial sites can be replaced by remote data surveillance techniques. This helps reduce trial costs and times and so contributes to closing T2. The adherence issue in T3 can also be addressed digitally. Not only can automated alerts be issued to keep patients adherent to their regimens, patient portals can be customized for the nature of any 'mental block' they may have—in terms of concerns about treatment or lack of sense of need.

Ultimately the goal of digital health should be to help integrate healthcare fully into normal life. It is striking that we think about keeping well and falling sick in two distinct compartments of our mind. The first is the stuff of magazines, diets, exercise regimens, mindfulness—an endless list of things for which we take personal responsibility and are more than willing to reach into our own pockets to fund. The data that control this part of our health life is on our smartphone.

Falling sick and entering the health system is in a completely different mental compartment. We are passive, dependent on the professionals, wishing we were anywhere but in the doctor's surgery or hospital. We expect the health plan or government-funded system to pay for all we need. And the data that leads to our diagnosis and results from our treatment is controlled by the system.

Digital health has the potential to bridge between these two worlds and, along the way, help us to think and act more holistically about our health lives. There have already been studies of physicians receiving a report derived from patient-generated health data from mobile phones. The majority of providers were willing to use the reports; a few said it actually changed their diagnosis[21].

All of this digital power and possibility raises the question: what about the new IT-based industrial entrants that are clearly beginning to eye the healthcare world as a new marketplace? There will be a host of start-ups designing apps and software modules, but the most intriguing new players are those that can apply huge investment and skills bases in totally new ways: Google, Apple, Amazon, Qualcomm, and the rest—mostly based in California, but with global ambitions.

Their success will not be automatic. They are not immune to the regulatory, reimbursement, and adoption challenges that face established healthcare players. But we already see the launch of useful tools, like Apple's Research Kit that can transform the productivity of consumer-based research[22]. This tool can reach out to populations with smartphones not normally involved in clinical research. Half the world's Parkinson's disease patients live in China: with its high mobile phone penetration, we can conduct larger, more detailed real-time epidemiological studies on such diseases than ever before. It can track tremor, balance and gait, speech characteristics, and memory, both before and after medication.

As healthcare goes digital and the 'digital natives', now fit and healthy, themselves become significant healthcare consumers, the greater the impact of digital health is likely to become.

Big Data

As we discussed before, 'Big Data' is not automatically more important than quality data. However, there is no doubt that scale enhances the value of certain data, especially in conditions in which natural variability and confounding factors make

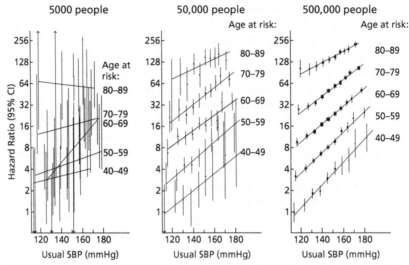

Fig. 7.2 Large numbers give clarity: ischaemic heart diseases vs. systolic blood pressure. Reproduced by kind permission of Sarah Lewington, on behalf of the Prospective Studies Collaboration, Clinical Trial Service Unit, Nuffield Department of Population Health, University of Oxford. Copyright © 2016 CTSU.

it difficult to separate signal (what's significant) from noise (what isn't). Scale can come from the number of patients sampled, from the range of exposures or outcomes seen, from the frequency of observation, and in the level of detail with which it is collected.

The benefit of scale is dramatically illustrated in Figure 7.2, kindly provided by Sarah Lewington of the Prospective Studies Collaboration. Increasingly large sample sizes are examined for the risk of ischaemic heart disease as a function of blood pressure. The correlation clarifies with scale: a link that is barely glimpsed on 5000 people comes into sharp relief with 500 000 (Figure 7.2).

Artificial intelligence and machine learning

The new tools of searching, manipulating, and correlating data to produce insightful answers from large bodies of information will ultimately have a huge impact on life sciences and healthcare.

There are at least three areas now being intensively explored—mining the biomedical literature, connecting it to treatment choice, and gaining new insights from the healthcare experience base.

As we saw in Chapter 2, the medical literature is now a huge beast. Hundreds of journals, with hundreds of thousands of papers appearing annually, contain a stream of research findings that no individual professional can absorb, even if they restrict themselves to a very narrow field of interest. Take colon cancer: in the 18 months between January 2014 and June 2015 there were 29 800 papers that dealt with the topic, of which 18 300 included genomic analysis. Of course, one could read only the highest impact journals, or the output of a defined list of leading researchers, but then one might miss, for example, a review article in *Cell Host & Microbe* exploring the links between our emerging understanding of the microbiome and the incidence of colon cancer[23].

Ideally what researchers and clinicians need is a digest of the literature relating to their topic, scored, and weighted by relevance (and perhaps reliability). Based on this review, they would ideally also receive a frequently refreshed depiction of the relationships between genes, proteins, pathways, and their interaction with drugs, as they are emerging from diverse research findings. They would like to know the strength of the evidence and so the robustness of the conclusions the authors drew. They would also want to know whether these findings aligned with past consensus, or whether they contradicted it—or found or concluded something totally novel. So we would want to have scanned not just the abstract but also the key sections of the paper itself.

This is an artificial intelligence problem, one that IBM has set out to solve using its 'Watson' methodology. It has also, with Watson Genomic Analytics, sought to marry the results of mining the literature on genomics, treatments, and outcomes with an individual patient's genomic analysis. It would go on to identify driver mutations, their implications for protein pathways in the tumour, and so the identification of 'treatable targets'. The next step, currently in the hands of clinicians (!) is the selection of the appropriate drug regimen. The early work by IBM, in partnership with Memorial Sloan Kettering Cancer Center, has been on glioma, since expanded to a network of US medical centres, but their ambition is to go much further—in terms of diseases and geography.

A third major challenge calling out for artificial intelligence approaches is mining the growing databases of medical record and personal health data from social media, the use of apps, and wearables. Artificial intelligence is already in use in predicting the propensity to develop a disease, or utilize a service. It is also effective at the detection of medical fraud. However the most intriguing artificial intelligence application from medical records and personal data will be the discovery and validation of unexpected correlations. These could relate to an off-target protective or therapeutic effect of a drug, or activities that bring on conditions[24].

So the technologies in the pipeline are potentially quite disruptive. Omics analyses that specify more precisely our diseases, imaging that detects and monitors with unprecedented levels of definition, advanced therapies with curative potential, wearable devices that give real time feedback on our condition, and digital applications that support a patient–doctor partnership in care as never before. And artificial intelligence tools whose long-term impact we can only guess at.

But it is the convergence of these technologies that will create the greatest leaps in our capability. When we bring together 'Dx', 'Rx', and 'Ix'—more refined diagnostics, new therapy modalities, and the right information tools to specify, monitor, and track them—we will have 21st century precision medicine.

We now go on to examine early evidence of the potential of these advanced tools, from genomics to digital sensors, coupled with new approaches to the innovation process, to address unconquered therapeutic challenges. Cancer, rare diseases, the prevention of chronic conditions, and the diseases of aging—especially dementia—need to be high up our priority list.

Cancer treatment

Precision medicine is poised to literally revolutionize cancer treatment.

Called the 'Emperor of All Maladies' for good reason, cancer is insidious, often invisible and undetected for many years, but then often spreads through the body apparently unstoppably. A 'war against cancer' was declared, as is the way with American presidents, by Richard Nixon as long ago as 1971. But, over most of the intervening years, the treatment paradigm has remained, essentially: detection by the appearance of a lump or suspicious pain; confirmation by a CT and/or MRI scan; surgery and/or radiation in an attempt to eliminate or reduce the tumour; application of chemotherapy based on cytotoxic drugs to mop up the remaining cells and reduce the probability of re-occurrence. Typically the chemotherapy regimen begins with relatively broad-based drugs that target DNA in rapidly dividing cells.

If the cancer is found to have metastasized, more specific regimens, more recently developed, come into play. These often contain an antibody that targets cell signalling pathways or specific antigens on the surface of the cells. If this is not ultimately successful, other combinations (of increasingly specific and expensive drugs) may be tried. In other words, try the simple and inexpensive and—if this fails—move on to the targeted and expensive.

The way in which trials are done, and approvals granted has reinforced this sequence of events. It has been seen as unethical to test the newer, more specific drugs on patients before they have failed on more normal regimen(s). This has resulted in many new products coming to regulators (and reimbursement agencies) with data showing a few months of life extension on a proportion of refractory cases. This is a tough test for new therapies, as by this point there are often multiple colonies of highly mutated and rapidly dividing cells, with little clarity on which will ultimately prove fatal.

With the progress in genomics, other omics, and new generations of anti-cancer drugs, this pattern increasingly seems anachronistic. It is certainly not in the interests of patients where the cancer is known to be relatively resistant to typical regimens (as we saw earlier in metastatic melanoma). Cancer treatment will become more targeted from the earliest stage of diagnosis.

Early detection will in part come from more sensitive imaging. (Perhaps the day will come when full body MRI imaging each year will detect the earliest stages of tumour formation.) Blood tests to detect circulating cancer cells are also under development and could form part of annual health checks at an appropriate age. Early detection is a very large part of the overall battle against cancer. Much of the difference in cancer outcomes (measured by 5-year survival) between countries lies in the effectiveness of screening and the speed of follow-up.

But the most exciting aspect of cancer management today is the flood of new therapies and targeting technologies coming into use. Not only do we have literally hundreds of 'designer' drugs in the pipeline, we now see immunotherapy coming into its own. We also see the expanding use of biomarkers as diagnostic tests on tumours that identify the exact molecular nature of the disease, and their more precise measurement through imaging, with its ability to quantify the response to treatments.

Panels of tests for the common mutations (e.g. *EGFR, KRAS* and *ALK* in lung cancer) are already available, both in hospital labs and via commercial testing services. However, in most major medical centres, especially in the USA, these are rapidly being superseded by WGS.

We have long realized that the patient's immune system could in principle deal with cancer as it does other invaders. But we have also discovered that the tumour cells quickly become adept at evading control. The immune system alone either fails to recognize the cancer cells or—as is the case for some solid tumours—recognize them and accumulate around the tumour yet seem powerless to kill it.

Two of the most exciting approaches we have discovered in the last very few years give the immune system vital assistance and seem capable of putting some patients into full remission. 'Checkpoint inhibitors' intervene in the mechanism by which cancer cells evade the body's immune responses. Checkpoint inhibitors are so-called because they check over-activation of the immune system, for example in response to infection; cancer cells exploit these to suppress immune response to their presence.

Three sub-classes of therapeutic antibodies have been developed: those that bind to the programmed cell death (PD-1) antigen on the body's T- and B-cells, to its tumour cell ligand (PDL-1), or to antigen 4 receptors on the surface of tumour-specific T-cells. Checkpoint inhibitors work best alongside other drugs in the regimen that help control cancer cell growth[25].

Depending on the cancer, with lung cancer and melanoma being most studied so far, the response rates are between approximately one-third and one-half of patients. This is much higher than is typical for therapies for such less tractable cancers (around 10%). We may see other checkpoint inhibitor approaches emerge, since the field is still relatively young.

The second major weapon that has joined the cancer arsenal in the last few years is the ability to engineer the patient's T-cells themselves to more powerfully and specifically recognize tumour cells in various haematological cancers. They are collected from the patient's blood, engineered to produce receptors on their surface called chimeric antigen receptors (CARs) that recognize the tumour cells. These cells are then grown in the laboratory and reinjected into the patient. Remarkable results have been achieved in patients with advanced acute lymphoblastic leukemia (ALL) and lymphoma[26].

The initial 'CAR-T' developments are obviously just the start of a journey of engineering heightened immune response. A recently launched UK company, Immunocore, is linking engineered T-cells with specific antibody fragments that alert the immune system to destroy the cancer cells. Early trials on metastatic melanoma are promising.

The scientific prospect is of exquisitely tailored drug and/or immunotherapy regimens that are designed around an individual's tumour—the much-heralded 'personalized' medicine. This 'n=1'precision is what will enable us to turn the normal sequence of treatment upside-down. We will have the confidence to give the highly targeted drug combinations at the outset, in so-called first-line treatment, immediately after surgical excision. And we will be able to use imaging tools, both PET and dynamic contrast MRI, and analysis of circulating DNA from apoptotic cancer cells, to check that the regimen is having the desired effect.

It will be a challenge to the development and regulatory process to arrive at this revolution. It should happen first in those cancers most resistant to conventional treatment, such as pancreatic cancer or metastatic melanoma, where patients should not be asked to try conventional treatments before they go on to the more advanced regimens.

Of course, these developments pose non-scientific problems that must be tackled before they can enjoy widespread use. First, we need to design clinical trials that recognize these genomic and other differences. The small numbers involved can make it hard to accumulate enough patients that share exactly the same patterns of mutation. In some cases, especially where one or both of the agents have established their safety profile elsewhere, we may be able to dispense with major combination trials altogether.

The second problem is reimbursement. While the smaller—and sometimes shorter—clinical trials will reduce development costs, they will still be expensive therapies addressing small markets. On a patient-by-patient basis, therefore, prices will seem high. Payers' concerns will be heightened by the possibility that doctors will be pressured to prescribe the therapies beyond the patient group whose profiles match the drug, especially where no other effective therapies yet exist.

These problems aside, this is a very promising time for cancer therapeutics, firmly in the precision medicine domain. The challenge will be to utilize the adaptive approach in a manner that brings the successive waves of innovation rapidly and affordably to patients, with the optimum use of targeting technology to ensure that the potential of these agents (and expenditure on them) is not wasted.

Rare inherited diseases

These are, of course, some of the most tragic of the health problems mankind encounters, when a child unexpectedly is born with a mystery condition that affects their development and also, in many cases, shortens their lives dramatically.

There are estimated to be around 7000 rare genetic diseases, so although each is rare, the cumulative incidence is remarkably high, in between 3 and 6% of newborns. As we saw earlier, in some societies, such as Saudi Arabia, levels are even higher, because of a greater level of intermarriage between cousins. This has led to many genetic conditions being first recognized in the Kingdom. Testing has begun against 13 gene panels covering all known Mendelian diseases and variants and first results are emerging[27].

One of the most challenging aspects for parents with children with inherited disorders is that, because of the rarity of each individual condition, the chances that their doctor, or even the local hospital, can recognize the disease is small. So they began what has been called the 'diagnostic odyssey' in which—sometimes over several years—they go from specialist to specialist until someone knows what tests to run to identify the disease. Of course, diagnosis does not immediately imply therapy, but even the certainty, the ability to put a name to the disease, is of great value to the family.

Genomic medicine has the potential to dramatically shorten the diagnostic odyssey and also the path to treatment, where one is available. In most major medical centres, exome (or even whole genome) sequencing is being implemented for rare inherited disorders. The patients don't need to see someone from among the small group of specialists in their particular syndrome, as long as their clinical geneticist has the results of the sequencing available and can initiate whatever confirmatory tests or therapies that fit the genomic diagnosis.

This area of medicine evokes some of the strongest reactions from both professionals and parents anxious to do the best for their children. There are many patient organizations, each devoted to a single disease, and 'umbrella' groups to unite their interests. Two UK examples of this are the Genetics Alliance and Findacure, and in the USA, the Genetic Alliance. Both 'Genetic Alliance' organizations advocate for the field and for strong engagement of patients and families in treatment choice, research, etc. Findacure makes a very important point in describing themselves as 'The Fundamental Diseases Partnership'. They remind their supporters that rare diseases are actually tremendously important windows into fundamental biology, the link between our genes and their role, with much wider implications.

So this area of medicine will see significant advances from the application of genomics, but it also deserves support from the general biomedical standpoint.

Avoiding chronic disease

Until now, we have regarded chronic disease as something that emerges, often in the middle years, somewhat 'out of the blue'. Of course, we are aware of risk factors—blood pressure and lipids for cardiovascular disease, obesity for diabetes, and so on. Family history has also provided pointers on what we each should look out for, but otherwise it is just the appearance of symptoms that alerts us that we have joined the majority of those in the second half of life with some form of life-limiting chronic condition.

But we now have markers that, in real time, track the underlying changes that precede any disease becoming symptomatic. Examples include the ability of Health Diagnostic Laboratory's panel of proteins and metabolites to identify pre-diabetes and Exact Sciences' test that indicates adenomas en route to colon cancer.

However, we can probe deeper and so probe earlier. The work of Lee Hood's Institute of Systems Biology on 107 healthy individuals shows that we can pick up the very first signs of potential disease through a mix of omics tests. Then, as in the new Arivale business model, health counsellors can step in, explain the significance of the results to the 'pre-patient' and guide them to the lifestyle changes that will postpone, perhaps for ever, the development of a debilitating condition.

Initiatives like this span the wellness/disease divide we talked about earlier in relation to health information. It extends the lifestyle-driven wellness management for which increasing numbers of the middle aged are prepared to invest substantial time and money. How else to explain the hordes of greying cyclists in colourful Spandex toiling up Mont Ventoux, the toughest part of the Tour de France.

In time, though, if the methodology can be simplified and made generally affordable, it will be seen as in the interest of healthcare payers to track their plan members and intervene before, not after, symptoms appear. One is reminded of the film, Minority Report, in which potential criminals were arrested before they committed a crime!

Diseases of aging

There are massive healthcare implications of the aging of the population. While in itself good news, the rapid growth of those above 65 will bring with it a huge increase in the diseases of aging: dementia, multiple chronic disease morbidity, and physical frailty. We are about to see a major increase in this proportion as a result of better healthcare combined with underlying population growth.

There is an increasing number of people that consider aging itself as a disease. There is a growing movement, typified by the SENS Foundation, that argues that many, if not all, of the biological changes that occur as we age are themselves addressable. If we were to do so, then many of the diseases that aging appears inevitably to bring would be postponed or avoided. Others argue that aging is biologically pre-programmed and we intervene at our peril. Yet others say that it doesn't matter whether or not aging is 'intended' by biology (to make room for the next generation), it can still be understood at the molecular level and its effects counteracted.

The process of cell senescence has received a lot of scientific attention. It is clear that molecular and supra-molecular damage accumulates, and the result is a range of age-related diseases, ranging from atherosclerosis to dementia. In fact, the correlation of many diseases with the 'cardiovascular age' of an individual suggests the underlying molecular mechanisms can be slowed, if not reversed. Sub-clinical inflammation may result from up-regulated cytokines and increase in oxidative agents in the body.

There seems to be a link to the decline in metabolic activity that occurs over time, and the shift from burning carbohydrates to burning fat. A lot of attention has been paid to whether caloric restriction can slow the process of aging, as it certainly does the development of Type 2 diabetes. Encouraging, at least for some, is the fact that resveratrol, best known as a component of red wine, has been shown to mimic certain beneficial aspects of caloric restriction, including protection against Type 2 diabetes, perhaps by inhibiting phosphodiesterases[28].

One clear feature of the aging cell is the shortening of telomeres—the thousands of repeat sequences of the bases TTAGGG at the end of the genome that shorten with each cell division and therefore with the passing of the years. A link has been established between telomere length and such symptomatically diverse diseases as liver cirrhosis and pulmonary fibrosis[29]. Mutant telomere-related genes are also associated with premature aging disorders, such as Hoyeraal–Hreidarsson syndrome.

At the molecular level, senescent cells with shortened telomeres secrete higher levels of cytokines and proteases. Can this process be stopped—or even reversed? We know that the enzyme telomerase can carry out such re-addition of the lost sequences. However, the concern is that just stimulating of this activity will awaken or promote the growth of cancer cells, increasingly likely to be present as the body ages.

There are less interventional approaches. Elizabeth Blackburn, the Nobel prize winning discoverer of telomeres, has demonstrated that meditation to reduce stress actually has a measureable positive impact on telemere length![30]

DNA repair is another promising avenue for countering aging. Human biology has a range of mechanisms for repairing the mutational damage that accumulates with cell division and aging. As often is the case, clues to these mechanisms are found in patients with accelerated versions of a more common condition (in this case the universal condition of aging). Accelerated aging in a 15-year-old boy was studied at Erasmus University; he was found to have a mutation in the *XPF* gene, which has a role in the nucleotide excision repair pathway. This pathway involves more than 30 different proteins, and its breakdown leads to faster accumulation of mutations, whether in the skin or the brain[31].

Other changes clearly occur as the biological clock ticks on. The immune system leaves the homeostatic state in which it has existed in previous years and either fails to combat intruders or gives rise to sub-clinical levels of inflammation. This may then spark rheumatoid disease, atherosclerosis, or even perhaps dementia.

We need to know more about such changes—a T0 problem. What can we learn about gene mutations, gene expression changes, protein levels, and so on that might give rise in one person to arthritis, in another heart disease, and possibly in a third some forms of dementia? We get some clues from the way in which diseases cluster. I recently met a lady with advanced RA, who had recently developed a bad continuous cough, a sign of interstitial lung disease. Going to the literature, I found that some form of lung deficiency, like pulmonary fibrosis, occurs in about a third of patients with RA. Can we map the changes—probably to the immune system—that links and perhaps underlies these two apparently very different diseases?

From the therapeutic standpoint, there are at least three approaches we can take: (1) stop or reverse senescence itself, (2) persuade senescent cells to die to make way for younger tissue, and (3) replace or supplement aging cells with transplanted cells.

The area of combating aging has drawn the attention of several prominent investors, including Google (via the new venture Calico) and a consortium of companies and foundations in Human Longevity Inc, founded by Craig Venter, who played a crucial role in the first sequencing of the human genome.

As these insights accumulate, so will possible targets for pharmaceutical intervention, as well as means of modulating aging via nutrition and other aspects of lifestyle. However, how will we be able to study their effects systematically and scientifically via clinical trials? What end points will we measure? How will we distinguish the effects of people with very different genomic profiles, whose aging processes may be going at differing speeds?

If we can intervene effectively in aging, with the multiple morbidities it currently brings, the impact we will make on the sustainability of the health system will be huge. The utilization of healthcare increases rapidly as we age through the 70s and 80s, and beyond. We all know of individuals who are in and out of hospital, appear frequently at their doctor's surgery, and never seem to be well.

But we are becoming more insightful at spotting patients who are on course for such a fate, using data analytics, and it will be very cost-effective to intervene before they reach a stage of illness or frailty at which their problems multiply. And the increasing trend to commission care for the elderly as a group, rather than just the episodes of care they eventually need, will accelerate those kinds of interventions.

Dementia

It seems appropriate to come to dementia after considering the molecular effects of aging, since dementia is now the most-feared age-related condition. As we saw in

Chapter 4, it threatens to overwhelm the world's healthcare systems, as demographics put more and more of us in its path.

Dementia and Alzheimer's Disease (AD) are terms that are often used interchangeably in the lay press. However, distinguishing between them, and indeed between different forms of AD is important if we are to untangle their molecular mechanisms. Dementia and AD represent one of the most important fields for precision or personalized medicine, since the disease seems to take a somewhat different course in each individual.

As we saw earlier, tangles and plaques of beta-amyloid protein are the most evident signs of the progress of AD, and have been the most common targets for early attempts at disease-modifying pharmaceutical intervention, but with—until very recently—uniformly disappointing results. It seems either that the misfolded protein is a mere side-effect of the development of the disease, or that we have yet to intervene early enough in its accumulation to make a real difference to patients' symptoms.

However, we can now begin to piece together some insights and tools that promise to transform our understanding of the disease and our ability to treat and manage it. These include: (1) factors that predispose to the condition, (2) diagnostic tests for early detection, (3) new routes to therapy, (4) increasing collaborative efforts, (5) imaging technologies to confirm diagnosis and track response to treatment, (6) regulatory innovation, including the use of adaptive development pathways, and (7) IT tools to help maintain independence and self-management in the earlier stages of the disease.

It is a mistake to regard dementia as a 'bolt from the blue', incapable of being anticipated, delayed, or avoided. In fact, dementia represents a fruitful area for exploring the power of Big Health Data. As we analyse such data we will find genetic and phenotypic factors, as well as environmental factors that contribute to disease risk and progression.

The factors that predispose to AD are both genetic and behavioural. Genomic analysis reveals that several genes, including the *ApoE* gene, play a role (this gene is one of those highlighted in the consumer-focused 23andMe test). There are three forms of the gene—*E2, E3,* and *E4* and there are therefore six possible combinations for each individual, with *E4/E4* the 'worst', conferring a ten-fold increase in probability of developing the disease.

Other genes have been implicated, including *CLU, PICALM, CR1, BIN1, ABCA7, MS4A, CD33, EPHA1,* and *CD2AP*. A transatlantic consortium (IGAP) is now investigating both the genes that increase likelihood and those that might confer some protection, by examining the genomes of 20 000 AD patients and 20 000 healthy controls[32].

There is also growing understanding of the *behavioural factors that increase risk*. Most of these are the same factors that increase risk of heart disease and stroke: namely the various elements of vascular health—blood pressure, blood lipids. The recently published FINGER trial showed that people in the 60–77 age bracket benefited from a series of interventions, including management of cardiovascular risk factors, physical exercise and social activities, and cognitive training versus a control group simply given regular health advice. Their memory, executive function, and speed of cognitive processing were all superior[33].

Interestingly, the most recent data on disease prevalence in the UK suggests that the age-adjusted level of dementia is actually falling[34]. If the finding is confirmed this could confirm that greater attention to optimum health earlier in life can affect cognitive decline later.

As we saw in Chapter 4, the search for effective therapies for AD has been a long and largely discouraging story. Most unsuccessful efforts have been following the clearest molecular pathology clues, in terms of circulating a-beta protein and aggregation of beta-amyloid in the brain. Just when the 'amyloid hypothesis' pursued by many companies seemed discredited, it was recently reported that one antibody, aducanumab from BiogenIDEC, showed not only reduction of amyloid plaques in the brain, but also positive impact on the decline in memory and thinking ability. This will undoubtedly encourage other companies to continue their efforts.

However, we still need *new routes to therapy* that build on other steps in the chain of events that lead to neuronal death. Phosphorylated tau is one of these. Another yet untested possibility is the use of cell therapy (based on stem cells differentiated into neurons) to replace brain tissue. There are some promising results in Parkinson's disease, but compensating for the loss of neurons across the brain, with their memory-bearing connections, may be beyond its reach. Replacement of cells in the hippocampus, those that help lay down fresh memories may be a different matter[35] but the invasive nature of the procedure may make it infeasible.

The other encouraging feature of the scene in AD clinical research is the increasing willingness for both medical centres and companies to *collaborate in trials*. A large-scale US consortium is working together to test interventions that are not industry-sponsored[36]. An international initiative between the USA and Europe's IMI promises to create a clinical trial platform that could cut 2 years off development times. This is an initiative supported by Johnson and Johnson, Eli Lilly and Company, Lundbeck, and Takeda[37].

So, although we still lack the promising pipeline of molecules we need, the innovation process is exhibiting the right features for rapid development once they appear in greater numbers.

Imaging already plays a very substantial role in diagnosis, and the development of PET imaging agents by several companies is enabling earlier detection of AD. We need to understand better the links between brain morphology and the early stages of disease, since some patients show plaques where there is little or no sign of cognitive loss, and vice versa. If we can develop more precise imaging markers, this will help identify early stage patients suitable for clinical trials. Imaging also can be used to track progress of the disease and monitor how effectively future therapy is keeping it at bay.

Regulators are actively collaborating with each other and with sponsors to challenge past ways of developing products. A single scientific advice route, greater data sharing, more consistent guidelines, modeling the use of surrogate endpoints, and adaptive development pathways are being investigated as part of the Global Dementia programme. It may be that dementia becomes the first area in which a truly global regulatory regime emerges. Raj Long, in her report to the Global Dementia programme proposed an International Dementia Advisory Platform in which public and private efforts receive screening and advice from a global pool of experts[38].

Inevitably, however, many patients will develop increasing levels of cognitive decline and it becomes vital to help them stay independent as long as possible and manage their lives to the fullest extent possible. So dementia is also a target for digital health. Several *tools for keeping the brain sharp* are now available, as well as the well-publicized techniques of games, language learning, and other ways of stimulating mental activity. In addition, developments like MyBrainBook, a UK innovation that enables patients to manage and track their lives, have an important place in helping patients continue to play their roles in family and society[39].

At later stages, personal safety and ability to take basic care of oneself become the priorities. Here 'intelligent homes' with sensors and video links will extend the length of safely independent life for dementia sufferers. Simple sensors can be useful, like those that report to family or a remote monitoring service if someone has switched on a kettle in the morning, or automatic gas sensors that alarm if unlit gas is escaping. The 'Internet of Things' will extend its reach into old age.

So, although we are far from seeing a dementia cure, the field will benefit from many of the seven steps from Chapter 5: a greater investment in molecular and systems understanding of disease, an adaptive mindset in development, greater levels of collaboration, and a precision medicine approach to therapy and patient support.

We now leave the specifics of the technologies in the pipeline and the diseases they might address to estimate the overall impact we should be looking to make in the productivity of the innovation process.

Assessing the impact

It would be a bold futurist who would forecast the overall impact of these technologies and changes to the innovation process on human disease and patient benefit. Let me illustrate it, though, by the changes being advocated in pharmaceutical innovation.

We have seen that gaps need to be bridged at the T0, T1, T2, and T3 stages, if a potential medicine is to target disease in a relevant manner, escape from the academic lab in the form of a clinical candidate, surmount the regulatory and reimbursement hurdles, get widely adopted by the clinical community and become effectively used by patients.

We have also seen the gross inefficiencies at each stage:

T0: many efforts fail because we do not have sufficient grasp of the molecular basis of disease

T1: too few researchers focus on the potential for translation and too much research is fundamentally untranslatable

T2: there is substantial attrition at every stage in clinical trials and several approved products fail to achieve broad reimbursement

T3: it takes 10–15 years for a new product to be widely adopted, and in some cases patients still fail to adhere to the prescribed treatment

What might be the impact on these gaps of the changes we advocate? We can only make very broad estimates.

T0: Suppose we redirect more basic research to focus on collaborative efforts to discover disease mechanisms in areas of unmet need. We could then use this knowledge to design model systems on which new drugs could be tested. Could we increase the potential for translatable research by 20–30%?—plausible?

T1: With increased 'nudging' and better training of academics on the translation process, could more emerge from their labs that has therapeutic potential? (We need only look at the wide range of translation performance across universities to answer 'yes' to this question.) More, and more effective, pre-competitive collaborations could focus on validated biomarkers and other tools for better targeting. Could we increase the flow of products from research into the clinic by 30–40%? Again, this seems to me perfectly possible, indeed rather conservative.

T2: Could we, with more precision medicine, including better patient selection and adaptive development pathways, reduce the current level of attrition in the clinic by 40–50%. Could we bring these better targeted products to patients 3–4 years sooner? If we look at recent experience in regulatory approvals, we see clear evidence of these possibilities. And if we involve HTA agencies and payers at an earlier stage and define the evidence they need, before and after conditional approval, could we reduce the delay in achieving reimbursement by 1–2 years and the failure rate by 20–30%? These seem to me reasonable targets.

T3: Can we use the engineering approach to adoption—CDS tools, peer-to-peer education mechanisms, and better post-approval tracking—to shrink the average time to peak use of an innovation from 10 years to 5? Challenging, but for high impact products that should be our minimum goal. Can more personalized adherence support, including patient education, compliance tracking, and digital motivational tools, increase the proportion of patients that use the product effectively by 40–50%? There is evidence this can be done.

T4: What additional treatment options, better targeting approaches, and new applications for existing products can we distil from the health data we will steadily accumulate? What applications of mechine learning will point us in new directions?

Many may find the analysis I have given too speculative for their taste, and I understand that. But the power of precision medicine lies in the multiplication of the potential benefits through the process, and the achievement of this benefit substantially faster. The final (steady state) level of patient benefit achieved, were we to take the midpoint of each of the suggested ranges, is around six-fold. But, taking account of the faster time-to-patient, and assuming a product life from discovery to replacement of around 20 years, the net present patient benefit (to coin a term derived from finance) would be more than ten times. (Figure 7.3).

This analysis takes no account of the unpredictable but very likely advances emerging from closing T4 and learning better how what have can be better used and repurposed.

My purpose here is not to make specific predictions but to illustrate the scale of the potential benefits of change if we can implement my proposals. More than an order of magnitude in total patient benefit might be achieved were we to work together to reform the medical innovation process along precision medicine lines.

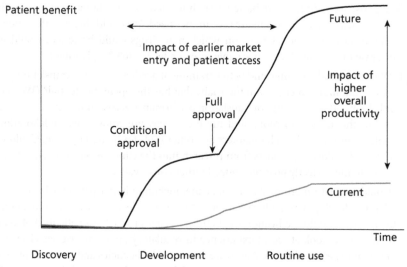

Fig. 7.3 Illustrative impact of change.

Summary of Chapter 7

To propel change forward we need not just a good sense of direction but also a sense of the prize, especially for patients, if we are successful. The technologies now coming into our hands have the potential to create precision medicine capable of overcoming some of mankind's most intractable challenges: cancer, inherited diseases, aging, dementia—among many others. Taken together, the changes we propose to the innovation process could bring at least an order of magnitude greater net patient benefit—over the lifetime of products that are more rapidly developed, better targeted, more consistently reimbursed, more swiftly adopted, and better utilized.

In Chapter 8, we explore what the life sciences ecosystem of the future ought to look like, what evolutionary forces will shape it, how the roles and relationships of its many participants are poised to change, and what new kinds of players may emerge.

References

1. The majority of Icelandic female settlers came from the British Isles, report on decode genetics website http://www.decode.com/the-majority-of-icelandic-female-settlers-came-from-the-british-isles/, accessed 01 Sep. 2015

2. Gudbjartsson DF, Helgason H, Gudjonsson SA., et al. 2015. Large-scale whole-genome sequencing of the Icelandic population. *Nature Genet*, 47:435–44.

3. Sulem P, Helgason H, Oddson A, et al. 2015. Identification of a large set of rare complete human knockouts. *Nature Genet*, 47:448–52.

4. **Jo Macfarlane,** Would you pay £4000 for Duran Duran's anti-wrinkle moisturizer? Article on MailOnline 2 May 2015 www.dailymail.co.uk/.../Would-pay-4-000-Duran-Duran-s-anti-wrinkle-moisturiser, accessed 01 Sep. 2015

5. **Barker RW, Brindley DA, Schuh A.** 2013. Establish good genomic practice to guide medicine forward. *Nature Med*, 19:530.

6. Clinical trials for stem cell therapies. Trounson A, Thakar RG, Lomax G, Gibbons D. 2011. *BMC Med*, 9:52.

7. Meeting highlights from the pharmacovigilance risk assessment committee (PRAC) 10–13 May 2016 http://www.ema.europa.eu/ema/index.jsp?curl=pages/news_and_events/news/2014/12/news_detail_002239.jsp&mid=WC0b01ac058004d5c1, accessed 01 Sep. 2015

8. **Cartier N, Hacein-Bey-Abina S, Bartholomae CC,** et al. 2012. Lentiviral hematopoietic cell gene therapy for X-linked adrenoleukodystrophy. *Methods Enzymol*, 507:187–98.

9. **Bowles D, McPhee SW, Li C,** et al. 2012. Phase 1 gene therapy for Duchenne muscular dystrophy using a translational optimized AAV vector. *Molecular Ther*, 20: 443–55.

10. **Sander JD, Joung JK.** 2014. CRISPR-Cas systems for editing, regulating and targeting genomes. *Nature Biotechnol*, 32:347–55.

11. **Hau PD, Lander ES, Zhang F.** 2014. Development and applications of CRISPR-Cas9 for genome engineering. *Cell* 157:1262–78.

12. **Wang H et al.** Multiplex automated genomic engineering (MAGE) report on Wyss Institute website, http://wyss.harvard.edu/viewpage/330, accessed 01 Sep. 2015

13. Wikipedia entry for Genome Engineering https://en.wikipedia.org/wiki/Genome_engineering, accessed 01 Sep. 2015

14. **Fakruddin M, Hossain Z, Afroz H.** 2012. Prospects and applications of nanobiotechnology: a medical perspective. *J Nanobiotechnol*, 10:31.

15. **Ballegaard C, Riis RG, Bliddal H,** et al. 2014. Knee pain and inflammation in the infrapatellar fat pad estimated by conventional and dynamic contrast-enhanced magnetic resonance imaging in obese patients with osteoarthritis: a cross-sectional study. *Osteoarthritis Cartilage*, 22:933–40.

16. **Frisoni GB, Fox NC, JackJr CR., Scheltens P, Thompson PM.** 2010. The clinical use of structural MRI in Alzheimer disease. *Nat Rev Neurol*, 6:67–77.

17. http://uk.businessinsider.com/science-of-elizabeth-holmes-theranos-2015-4?r=US&IR=T)

18. **Sarwat I. Chaudhry SI, Mattera JA, Curtis JP.** et al. 2010. Telemonitoring in patients with heart failure. *N Engl J Med*, 363:2301–09.

19. **Franzini L, Sail KR, Thomas EJ, Wueste L.** 2011. Costs and cost-effectiveness of a tele-icu program in six intensive care units in a large healthcare system. *J Crit Care*, 26: 329.e1–6.

20. www.aparito.co/

21. **Nundy S, Lu C-YE, Hogan P, Mishra A., Peek ME.** 2014. Using patient-generated health data from mobile technologies for diabetes self-management support: Provider perspectives from an academic medical center. *J Diabetes Sci Technol*, 8:74–82.

22. Empowering medical researchers, doctors and now you. Apple website page on ResearchKit and CareKit https://www.apple.com/researchkit/, accessed 01 Sep. 2015

23. **Sears CL, Garrett WS.** 2014. Microbes, microbiota, and colon cancer. *Cell Host & Microbe*, 15: 317–28.

24. **Banaee H, Ahmed MU, Loutfi A.** 2013. Data mining for wearable sensors in health monitoring systems: A review of recent trends and challenges. *Sensors*, 13:17472–500.

25. **Sharon E, Streicher H, Goncalves P, Chen HX.** 2014. *Chin J Cancer*, 33:434–44.

26. CAR T Cells in Acute Lymphoblastic Leukemia, Park JH and Brentjens RJ on The Hematologist website http://www.hematology.org/Thehematologist/Years-Best/3603.aspx, accessed 01 Sep. 2015

27. Research Scientists Release Initial Data from the Saudi Huamn Genome Program on the ThermoFisher website 19 May 2015, http://news.thermofisher.com/press-release/life-technologies/research-scientists-release-initial-data-saudi-human-genome-program, accessed 01 Sep. 2015

28. **Chung JH.** 2012. Using PDE inhibitors to harness the benefits of calorie restriction: lessons from resveratrol. *Aging* (Albany NY), 4:144–5.

29. **Armanios M.** 2009. Syndromes of telomere shortening. *Annu Rev Genomics Hum Genet*, 10:45–61.

30. Can meditation help prevent the effects of ageing, Jo Marchant published on BBC Future website 1 July 2014 http://www.bbc.com/future/story/20140701-can-meditation-delay-ageing, accessed 01 Sep. 2015

31. **Sijbers AM, van Voorst Vader PC, Snoek JW, Raams A, Jaspers NG, Kleijer WJ.** 1998. Homozygous R788W point mutation in the XPF gene of a patient with xeroderma pigmentosum and late-onset neurologic disease. *J Invest Dermatol*, 110:832–6.

32. EU-US collaborative study identifies 11 new loci for Alzheimer's, published on JPND research website 1 Nov. 2013 http://www.alz.org/news_and_events_21649.asp, accessed 01 Sep. 2015

33. **Kivipelto M, Solomon A, Ahtiluoto S.** 2013. The Finnish geriatric intervention study to prevent cognitive impairment and disability (FINGER): study design and progress. *Alzheimers Dement*, 9:657–65.

34. **Wu Y-T, Fratiglioni L, Matthews FE,** et al. 2015. Dementia in Western Europe: epidemiological evidence and implications for policy-making. *Lancet Neurol*, 15:116–24.

35. **Taylor CJ, Jhaveri DJ, Bartlett PF.** 2013. The therapeutic potential of endogenous hippocampal stem cells for the treatment of neurological disorders. *Front Cell Neurosci*, 7:5; ecollection 2013.

36. Alzheimer's disease cooperative study, Report on NIH website, http://www.nia.nih.gov/research/dn/alzheimers-disease-cooperative-study-adcs, accessed 01 Sep. 2015

37. **Ranit Schmeizer,** Global Alzheimer's platform and Innovative Medicines Initiative, 19 March 2015, http://www.ceoalzheimersinitiative.org/global-alzheimer's-platform-and-innovative-medicines-initiative-sign-memorandum-understanding, accessed 01 Sep. 2015

38. Dementia, report by the UK Department of Health, November 2013, https://www.gov.uk/government/uploads/system/Dementia_report.pdf, accessed 01 Sep. 2015

39. MyBrainBook website www.mybrainbook.com/, accessed 01 Sep. 2015

Chapter 8

Conclusion: New roles and relationships in a new innovation ecosystem

The ecosystem that has delivered medical innovation in the past is, by common consent, well past its 'sell by' date. In this chapter, we explore the forces at work reshaping it, and consider whether they will help or hinder the changes that we need in the innovation process.

In pharmaceuticals, the historical approach was one in which academic leads—for example, potential drug targets published in the scientific literature and presented at meetings—are picked up and pursued by several individual major pharmaceutical companies. In their laboratories, they then create lead molecules and produce many variants on the same molecular theme. The best of these are then taken separately by several companies through development and approval, and into marketing. Although there is the stimulus of competition, this also involves a great deal of duplication, and often multiple clinical RCTs. Sometimes all succeed; sometimes all fail.

This formula is already changing to some extent. Lately most innovative products have been initially discovered and developed by smaller biotech companies. The majors then vie for acquisition of the products or companies after first successful efficacy testing, usually after Phase 2. Indeed, if we examine the products approved by the FDA over the last 2–3 years we find the majority were discovered, and several brought to market, by companies from outside the top 20 pharmaceutical league.

However, even this modified process still has major inefficiencies. Some of the biotech companies run out of money before their products have completed Phase 2 trials, and such products are typically not picked up or further developed. There is still considerable duplication in clinical trials. And the whole enterprise will avoid tackling areas of high risk (like many psychiatric conditions) or heavy investment (like Alzheimer's).

How will the forces at work reshape this pattern of activity? Let us look at each participant and how their roles are changing. We will then go on to examine the changing relationships between them, in an era in which a greater level of collaboration and a more 'open' innovation culture are emerging.

Academia—a greater focus on impact

Academic research has long been the wellspring of life sciences. This will continue. But, as we have seen, the tendency for academics to focus only on their grants and publications has stifled the translation of their discoveries into genuine innovations. Too often researchers have lacked the necessary entrepreneurial spirit and a willingness to partner with a commercial company with the skills to build towards a real product. In many cases, the treadmill of grant applications simply leaves them with insufficient time to think about translation.

However funders, both public sector and research charities, are becoming more focused on the practical impact of the research they fund. Some are specifically funding translational research involving commercial partners. In the UK, 'impact' is an important measure of success that feeds directly into the university funding formula.

In the past, there have of course been academic–industry relationships, but the large majority of these have been sponsored research and individual appointments that leave the commercial sponsor on the side-lines of the research programme. There are now several new formulae being tried to create more productive partnerships, without hampering the investigative spirit of the academic. One is the 'embedding' of company staff alongside academics in the lab. A second is the support of fellowships for young researchers that spend part of their time in each environment. A third is the creation of consortia to conduct pre-competitive research whose direction is influenced by, and results visible to the industrial consortium. The SGC is a leading example of this model.

The UK leader of the SGC moved from a major company, GSK, to Oxford University. Around the same time, the current head of GSK's R&D organization moved from UCL to the company. These are good examples of the kind of two-way interchange that needs to occur between these two rather different worlds if their common innovation agenda is to advance. We need action on the incentive and career structures in both environments to encourage more of the same.

Most university technology transfer offices also need to change their approach in life sciences. Typically these bodies are tasked to maximize the licensing revenue from university inventions. This might be the right orientation in areas such as engineering in which a specific and clear-cut invention can be 'transferred' to a commercial company for a fee and a future royalty. However, in life sciences, these discrete inventions are few and far between. More often there is a promising lead, in terms of a drug target, a potential biomarker, or a new 'lead' molecule that requires collaborative work over years to turn into a commercial product—a diagnostic or a drug. Those guiding researchers from the university standpoint need to take a more long-term partnership perspective.

While university spin-outs will undoubtedly continue, there is a growing realization of the need for critical mass: more than a single technology or a single potential product. Accelerators or incubators will host them until they are externally fundable. So academia's role will inevitably become more innovation-focused and translational research will receive higher priority.

Industry—collaborating and partnering for success

It is now widely recognized that the 'fully integrated' model of pharmaceutical R&D— in which major international companies take promising leads from the literature, conferences, and academic collaborations, discover potential drugs in their own laboratories and move them through all the stages of clinical trial—has largely failed.

So the major companies have increasingly turned to acquisition of biotech SMEs to fuel their product pipelines. However, these are often very big bets: an alternative approach is to move away from such one-time acquisitions of previously unassociated companies to building an earlier, closer set of ties with several promising SMEs. It is becoming increasingly attractive for major companies to develop R&D relationships with a large family of biotech 'planets' allied with them. This enables the smaller companies to retain the unique innovative cultures that have made them more productive while the larger company has an option to license or acquire products or the company itself when the time is right. Figure 8.1 illustrates this phenomenon for Celgene and its family of relationships.

We need to go further in disease areas that will only progress when several large companies (and often other research funders) pool their resources. This is beginning to happen in diseases, like cancer, with a strong genomic component, as public and private sector investment generates 'open commons' databases on which all research enterprises can draw. The IT company, Intel, for example has partnered with Oregon Health & Science University to create a 'Collaborative Cancer Cloud'—a platform for institutions to share genomic, clinical, and imaging data[1].

Companies are also forming alliances to co-investigate their products in tough diseases in parallel, 'umbrella' trials. In such trials (e.g. the iSpy breast cancer trials), each company contributes one or more potential product and these are compared in different arms of the same trial. It is estimated that this approach can reduce the costs of development per successful agent by more than a third[2].

Patient groups and research charities—becoming drivers of innovation

The influence and activity of these organizations is growing. In the past they have acted as research funders to some of the most innovative research in their fields. Some, like the Michael J Fox Foundation and Cancer Research UK, have created their own R&D programmes, and they have effectively advocated for greater research spending and treatment support for 'their' diseases (although of course often within a situation that is a 'zero sum game' of research and healthcare budgets). There is clearly a new level of activism afoot.

Examples of this greater activism is a stronger focus on assessing the impact of the research they fund, more direct investment in disease-specific assets like databases and patient records and a more active role in searching for cures. Indeed, some frustrated with the slow progress in their field have incorporated 'cure' into their names (e.g. CureParkinsons in the UK).

In some cases, research charities have licensed molecules that are not progressing through pharma company pipelines and have taken their development forward.

They also have taken a role in the pricing debate, in some cases advocating an expensive treatment they see as revolutionary (such as Ivacaftor, see Chapter 4) but in other cases criticizing commercial pricing policies where they see them as excessive.

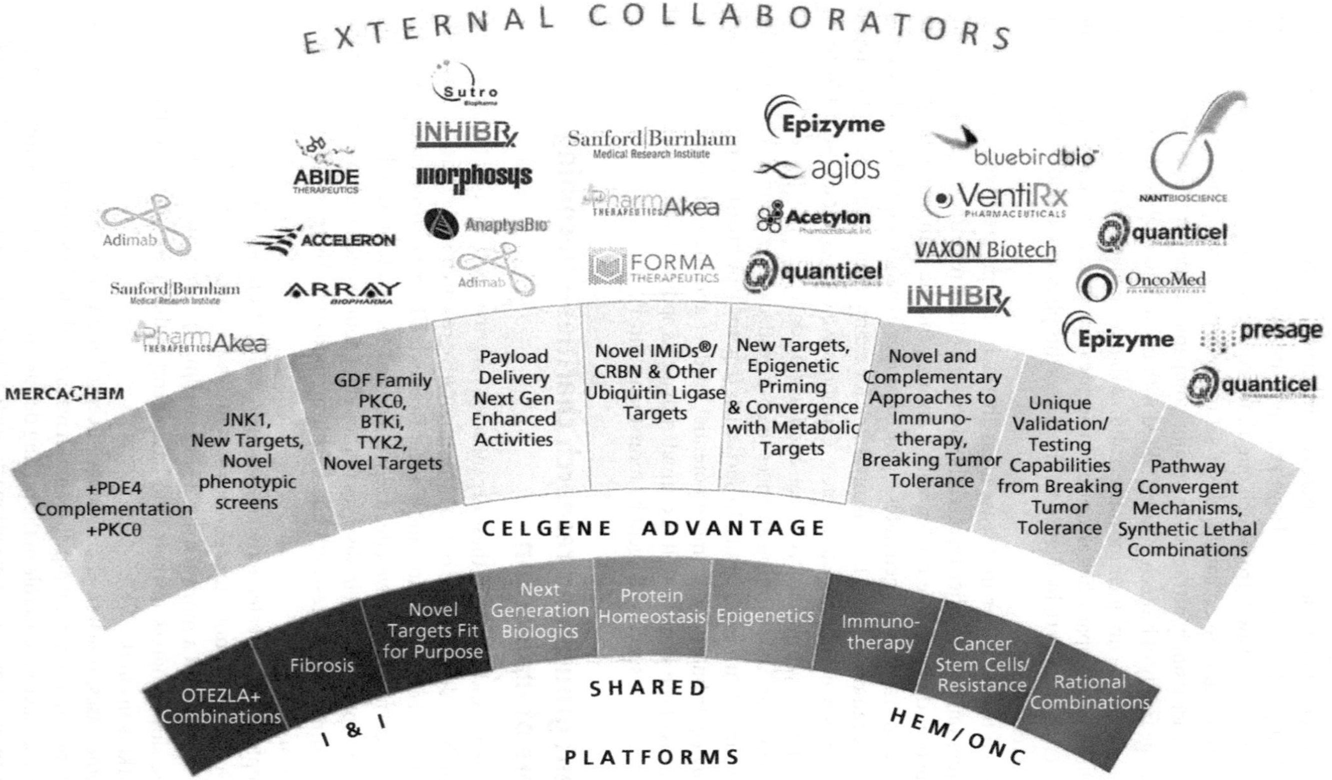

Fig. 8.1 Celgene external partnerships.

Reprodcued by kind permission of Celgene Coporation. Copyright © 2016 Celgene Corporation.

All the signs are that this activism will increase, as the patent/consumer voice strengthens in all sections of society.

Health systems—from passive customers to active developers

Of course, there are medical practitioners individually very engaged at every stage of the innovation process. And, in the case of devices, they are often the original inventors. However, the health systems within which they work have tended to be passive customers for products.

They will play a more proactive role in the future ecosystem, identifying the problems they want solved, becoming development partners in some cases, and being actively engaged in assessing the outcomes from their use.

It has long been assumed that the wisest choice of areas to research is best left to a combination of academic researchers, driven by the scientific interest and the emergence of tools to follow it, and commercial companies and their investors, determining where the best markets are likely to lie. As health systems face more and more pressing sustainability problems of their own, they will get more actively involved in setting research targets. Progressive health systems will also work with innovators to determine the evidence they want from clinical trials. Some may also be prepared to co-fund certain developments, for concessions on price.

Venture capitalists and the investor community—driving for near-term products?

Venture capital has gone through several cycles since it played a major role in the emergence of the biotechnology industry in the 1970s and 80s. In the initial phase, their patience was largely rewarded, while the shape of the companies they funded changed, sometimes beyond recognition.

Subsequent phases saw venture capitalists support 'platform' technologies, with the potential to create new generations of products, but that strategy became increasingly unpopular. At the same time the opportunity for funds to support standout successes in the IT sphere became more numerous. So the numbers of venture capitalists active in life sciences declined in the 2000s, with the vast majority remaining being located in the major US clusters. Most of these were really only interested in opportunities that led to real revenues within a few years. Several funded SMEs that were spin-outs formed around smaller products from major companies.

With the revival of the fortunes of the whole sector, funds are becoming more active again, and somewhat bolder. There is greater interest in molecular diagnostics and genomics. However, investors in drug discovery companies have seen the opportunity for profitable exits via large company acquisitions once a lead product reaches a successful Phase 2 outcome, and we may see fewer mid-sized companies emerge as independent majors.

It is interesting to see that much of the current 'venture capital' entering life sciences is the corporate and personal capital of the IT giants that have emerged in the Web era.

New entrants—ambitious and well-funded

Two broad groups of completely new entrants are entering the medical innovation fray. The first is a series of companies that focus on genomic or multi-omic data generation and bioscience analytics. These include NantHealth, Human Longevity Inc, Omicia, Cypher Genomics (now owned by HLI), and Foundation Medicine (now majority-owned by Roche). All of these are seeking to deduce new insights for research or treatment from analysing huge bodies of genomics data, sometimes coupled with phenotypic data such as proteomics or imaging. They realize that proprietary knowledge is more important and valuable than simply generating the data.

Of course, the companies that have created the technology that makes all this possible, like Illumina and ThermoFisher (formerly Life Technologies), are also integrating forwards to create added value solutions around the information that they help generate. This is all reminiscent of the strategies of computer companies in a previous era.

The second group of companies come from left field, in terms of technology. They are the IT giants, both the more established like IBM, Microsoft, Intel, and Oracle and also the upcoming 'Web 2.0' generation of Google, Apple, and Amazon. The former have a track record in healthcare, as suppliers of IT solutions. They are now creating new tools like IBM's Watson, that are aimed at applying new forms of intelligence to health data. Watson is already able to read the whole biomedical literature, weight its findings for reliability, and then link this information together around protein pathways implicated in disease and drug action. Oracle has a suite of eClinical solutions, building on its document management expertise.

The 2.0 healthcare IT companies are asking: what can our dominance in search, smart mobiles, and data analytics enable us to do that are beyond the resources or imagination of existing healthcare organizations? There are some intriguing links between these organizations, for example the hiring by Human Longevity of the creator of Google Translate. If a machine, they argue, can translate between two different languages neither of which the machine 'understands', can the same be possible between genotypic and phenotypic data? It would be a bold soul that would say this is a hopeless quest.

The outline of the radically different, and very intriguing, life sciences 'ecosystem' of the future is beginning to appear. Many of the driving forces described in this book have the potential to shape a more productive future (see Figure 8.2).

Most of the forces and the new relationships they result in should support the changes we need in the innovation process:

♦ Academia more focused on impact will invest in more translational capability and industry partnerships.

♦ Webs of collaboration between SMEs and larger companies will move investment into the urgent, nimble, and ambitious culture that SMEs offer.

♦ More activist patient organizations will also support translational research and invest directly in product development in a way that reflects patient needs more directly.

♦ New IT-based entrants will bring well-funded and disruptive inputs to the ecosystem.

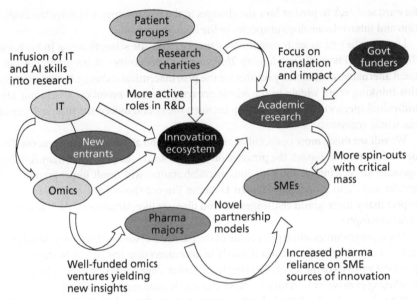

Fig. 8.2 Driving forces in the future medical innovation ecosystem.

It is not clear that these forces can reinvigorate the established pharmaceutical majors. Instead, we are creating a buoyant and promising SME sector and the real prospect of more high growth, mid-sized, independent companies like BiogenIDEC, Celgene, and Gilead emerging. These companies have become major players in the more specialist disease areas of multiple sclerosis, myeloma, and anti-virals. In these areas it is feasible to bring products to market without the huge Phase 3 trials, investments, and medical representative forces of most of the majors. However, they will need to retain their entrepreneurial cultures as they grow.

IT majors like Google and genomics-based ventures like NantHealth are pouring hundreds of millions of dollars into entirely new ways of combining bioscience insights, 'big data', artificial intelligence, and the outpouring of 'omics' information into products and services. While in many cases their own business models are unclear, they are offering partners, both private sector and academic, capabilities they could never have created. Data mining combined with machine learning will identify both new uses for existing products and new ways of stratifying the patient population for future products. Vast banks of genomic data, coupled with everything from electronic medical records to imaging—and coupled with machine learning—will yield insights into disease and new research avenues at the molecular level.

It is not necessary for all these new entrants to succeed in business terms for this new investment to provide a major fillip to life sciences R&D.

The term 'ecosystem' invites us to draw on the concepts and analogies of biology, such as evolution by mutation and natural selection. In his seminal book *The Origin of Species*, Darwin was looking back and explaining how, little by little, successful adaptations were responsible for the evolution of new species. Perhaps we can also look forward and seek to predict how the changes in the environment will drive the evolution and inter-relationships of species in the life science world.

I am not the first to apply evolutionary theory to the life sciences arena. In his book *The Future of Pharma: Evolutionary Threats and Opportunities*, Brian Smith has outlined alternative evolutionary futures for the pharmaceutical industry. But let us apply this thinking more widely to the whole ecosystem—since evolution affects not just individual species but the relationship between them, and indeed the sustainability of the whole ecosystem.

We will see much more open, collaborative innovation at the start of the process. The desire for tangible impact, the pressure on resources, and the premium many funding agencies are placing on inter-institution collaboration will result in programmes of greater scale and scope. The Human Genome Project showed the way, and we can expect many more 'grand challenges' around diseases like Alzheimer's, ALS, and muscular dystrophy.

These programmes will replace some of the patented research of industry, which will either receive the IP protection it needs to commercialize the products via creating practical variants of the leads emerging from earlier stage collaborations, or be granted marketing exclusivity in return for their clinical development investment.

We will see what I call the 'adaptive mindset' shape these clinical programmes. Not only will trial design be fitted to the nature of the product and need, but the blend of clinical RCTs and RWD data will also be customized to fit the purpose. And even clinical scale trials will increasingly be collaborations between companies, or between companies and patient charities.

As the role of charities in R&D becomes more intensive, this will drive consolidation among them, as supporters realize that just a few million dollars, euros, or pounds are unlikely to make a crucial difference to the global search for cures.

The majors themselves have an evolutionary problem. Fewer and fewer products are the kind of immediate 'blockbusters' that make a substantial difference to the revenues and profits of companies with tens of billions in revenue. They may have to accept products focused on a specific speciality, and then roll them out over several other niches where there are common disease mechanisms.

The diagnostics industry's growth will be driven by precision medicine, with a wide variety of genotypic and phenotypic tests coming forward. There will continue to be specific combinations of diagnostics and therapeutics, in which the diagnostic development and sometimes supply will often be funded by the pharmaceutical partner. However, increasingly there will be panels of tests and applications of whole genome and exome sequencing unconnected to specific drugs. Payers will need to find mechanisms to reimburse these, because of their impact on targeting expensive treatments. More money is moving into the diagnostic sector, and it is poised to take a more appropriate share in the value generated by precision medicine.

Major informatics companies will fund efforts to distil knowledge from ever-increasingly large and complex data sources. However, most of the smart applications will come from academics and start-ups, and the 'apps' they produce will be made widely available for modest cost, with the major companies benefiting from the overall volume of use of their platforms.

Many of these apps will enable patients themselves to understand and control their own health information, access education and choice about their disease, and work in partnership, with their clinicians as advisers rather than all-knowing decision-makers.

The interplay between the information part of the ecosystem and those engaged in other types of products will intensify. Mined healthcare information will produce key insights to direct research. Finding the right patients for clinical trials and treatment will become much more an information challenge than one dependent on physical activities. These are examples of value flowing from the product business into the information business.

As part of this phenomenon, I foresee the emergence of three categories of new information-focused companies:

+ Selection: matching patients to therapies
+ Adherence: supporting patients in their personal treatment plans
+ Outcomes: tracking the long term effectiveness and safety of treatments

The first will work with health systems, patient groups and product companies in implementing a precision medicine future: *finding the patients for targeted therapies*. Ongoing mistrust of the product companies, particularly pharma companies, will probably prevent them playing this role—so 'honest broker' intermediates may emerge. They would be rewarded in relation to the number of patients that enter a new therapy.

The second category could be *adherence support services*, again acting as intermediaries ensuring adherence to the therapy, taking into account the personalizing factors described earlier. They would be rewarded in proportion to demonstrated levels of adherence and persistence with treatment.

The third kind of information-empowered intermediary would be responsible for *tracking outcomes*. This will be increasingly important as product companies are themselves rewarded, at least in part, on outcomes rather than sales.

Obviously these three categories could eventually merge into single 'precision medicine service companies'. They would represent a neutral intermediary between pharma companies and patient-specific data. They would also require a skill set that is currently not within any of the other players. They could, of course, be funded by companies like Apple or Google, and thereby have access, with permission, to the rest of the patient's health data. This, of course, depends on the ongoing levels of trust these companies enjoy.

Since health budgets and medicines budgets are unlikely to rise to accommodate the new players, the service fees will probably need to be paid by the pharma companies. They will anyway need fewer medical representatives, if it is science rather than promotion that drives uptake.

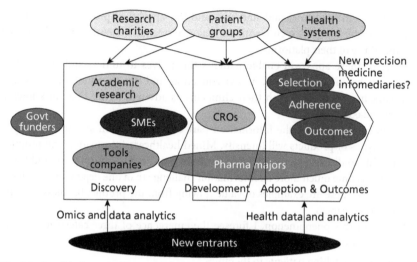

Fig. 8.3 Possible future ecosystem and roles along the value chain.

Patterns of competition and of collaboration will change. The cash flows of those commercializing major products will still enable them to construct a web of funded partnerships with academics and smaller companies. However, we will also see entrepreneurial health systems and governments—as well as IT giants—supporting technology developments and share in their success.

We can depict the future prospective roles of the various players along the value chain (Figure 8.3). The new precision medicine infomediaries just described will be mainly focused on the implementation stage. For some of the established players, like research funders, their role will change little. However, we will see important new inputs by patient groups at each stage, greater involvement of health systems in development, adoption, and outcomes measurement, and new interventions by the IT-rich players, both in discovery and in implementation.

If this new ecosystem has the potential to be more productive than today's, are there policies or incentives that can accelerate its arrival? Funders, government and foundations, can favour collaborative research models and the development of public–private partnerships to solve common problems. Universities and their funders can put greater emphasis on translation and partnership with the private sector. Patient organizations can use their collective voice to encourage a bold focus on unmet medical need. Health systems can engage in discussions with industry on development priorities, the funding of priority clinical research programmes and the kinds of creative reimbursement structures that address affordability without impeding new products and so represent win–win solutions.

So far this chapter has focused on pharmaceuticals. Box 8.1 contains some comments on how the ecosystems for diagnostics and devices may evolve.

Box 8.1 Future ecosystems for diagnostics and devices

Improving the efficiency of diagnostics development may need similarly radical changes, since the margins in this industry are not conducive to funding the kind of costly clinical utility trials that regulators and payers are increasingly asking for. Already we have partnerships between pharmaceutical and diagnostics companies to develop 'combination' diagnostics. However, as the use of targeting tests becomes more widespread we have the need to make available panels of tests that can be used to target drugs from several manufacturers. Who will pay for their development?

We may need less change in the devices ecosystem. Here small start-up companies have often grown up around the inventors, themselves often the surgeons who first experimented with using the devices. Once established, these companies are bought by one of the major international companies able to scale-up manufacturing and global marketing. This formula may continue to function well—at least until more complex intelligent devices are designed that are beyond the development resources of venture companies.

Conclusion

My final thesis is this. If we are bold enough to take the seven steps to sustainability and shape more positively the environment in which life sciences R&D takes place, then we can unleash forces that will lead us to a much more productive innovation ecosystem. At its heart will be precision medicine, broadly defined—using the new tools of biology, engineering, and informatics to match patients with therapies more accurately, and to track their outcomes more consistently.

Like any dynamic ecosystem, its future evolution will undoubtedly hold surprises, but it will make much less surprising and more probable the fictional story in Box 8.2.

Let me close with a challenge to readers from the different stakeholders that are critical to reforming medical innovation and ushering in the new era of precision medicine.

Patients

Press for and take your place at the decision-making table. Ensure you are well briefed on the development, regulatory, and reimbursement processes; you may not agree with all their stipulations, but you need to understand them. Be clear about the benefits you seek from innovation and how those should be measured. Forge relationships with leading academics who can support your position. Work with innovator companies, but avoid the risk of being seen as their mouthpieces. Help in the recruitment for trials. Take the lead in those issues, such as data ownership, access, confidentiality, and use in research, on which your voice should be decisive.

Make sure your sense of urgency communicates itself to other stakeholders when they are tempted to opt for conservatism.

Regulators

Use the flexibilities you have, and press for more. Be a 'critical friend' to high impact innovations, providing clear input on the evidence you need and the standard you will accept. Factor in patients' views of what represents benefit and what risks are acceptable.

HTA agencies

Examine your definitions of value to ensure they reflect both what matters to patients and what will help drive important new avenues of development. Be prepared to give early input on your evidence needs, so that they can be incorporated in development plans from the outset. Be prepared to issue conditional reimbursement guidance while confirmatory evidence is collected.

Payers

Recognize that, without innovation, the challenge of sustainable healthcare is unachievable. Take a full part in defining the evidence needed to purchase an innovation once developed. Support health systems in engineering much faster adoption and more appropriate utilization.

Clinicians

Recognize the legitimate voice of the patient. Ensure your clinical research is robust and unchallengeable. Play your part in engineering the adoption of innovation, as an opinion leader or workshop member, asking critical questions about evidence but then rapidly adapting your practice to incorporate value-proven new products.

Innovator companies

Be prepared to engage with all members of the customer community—including patients, clinicians, and payers—on their priorities for your R&D efforts. Approach development as an opportunity to meet their different needs. As you seek flexibility from others, show flexibility yourself in collaborative research and the design of clinical development. Be prepared for, and propose, novel forms of reward that help payers afford your innovations.

Politicians

Invest in understanding the realities—the investments, the timescale, and the global nature—of medical innovation. Support the experts in regulatory and HTA agencies in showing flexibility. Seek to shape an environment in which the needs of the health system and the opportunity to grow life science businesses come together, rather than stay in constant tension.

I recognize this is a tall order, for all those involved. We need to set 5- and 10-year goals for change and drive towards them. In 5 years we should aim at:

- Collaborative research consortia on unaddressed disease mechanisms
- Enhanced impact measures and incentives to help drive academic translation
- Adaptive and accelerated pathways in place for priority products
- Greater international convergence on evidence standards for efficacy, safety and value
- A global convention on patient data ownership, access and security
- Engineered processes for adoption of and adherence to high value products

  In 10 years our goals should be:

- Products for diseases like ALS and muscular dystrophy successfully treating patients
- New generations of antibiotics controlling anti-microbial resistance
- All cancer patients with their tumours genotyped and with personalized treatment regimens
- A single global evidence package accepted in all major jurisdictions
- Patient ownership and control of all personal health data
- The anonymized results of all treatments available for analysis

Embracing goals like these will create a future of truly sustainable innovation, one resting on the full power of precision medicine. This future is there to be grasped.

Box 8.2 A story set in the future, but not far in the future?

There had been no new therapy for ALS, otherwise known as motor neuron disease, for more than 20 years, and nothing that had any real impact on a wasting disease that typically took the lives of sufferers within 3–4 years from diagnosis.

There had been many attempts to try potential drugs, often not in formal trials, but nothing seemed to do more than give temporary respite. Lack of deep understanding of the molecular pathology of the disease, and a lack of effective animal models to use in preclinical development, had held back progress.

However a number of basic and applied research insights began to change that picture. Genomic and proteomic analysis of the sufferers from ALS, coupled with artificial intelligence—applied both to the data and to the literature—identified a number of genes that sat in two pathways of clear relevance for motor neuron function and the inflammatory response that could be triggering loss of function.

So this began to refine the search for drugs, both in terms of new targets for drug design and also the process of screening existing products for potential therapeutic value. Then researchers developed an experimental stem-cell-based system that allowed new drugs to be assessed in vitro.

(Continued)

Box 8.2 Continued

Because of the difficulty of the field and therefore its high risk as a commercial prospect, the three companies interested in it formed a consortium, along with leading US and European patient organizations and a major research charity. They met to discuss how best to fund research and measure the effectiveness of potential therapies at an early stage. The patients agreed that—given the very short and severe progress of the disease—they were willing to try any treatments with some prospect of working and to help organize patient registries and trials.

In parallel, other genomic research, coupled with imaging studies, had begun to identify subgroups of patients whose disease course and perhaps underlying molecular causes seemed different.

A year later, the consortium had found five prospective medicines, three of which were already approved for diseases that seem to share some of the molecular pathology that was emerging for a particular subgroup of ALS patients, who were identified by a combination of a particular panel of gene mutations and over-expression of some immune system proteins.

At this point, cross-stakeholder discussions were held, involving patients, clinicians, HTA agencies, and payers, to discuss what clinical and economic evidence should be sought in trials. While no guarantees of regulatory or reimbursement approval were given, there was a clear understanding of the metrics that regulators, HTA agencies, and payers would apply in their decision-making.

Following a short safety study for the two candidates that did not already have an established safety profile from existing use, the consortium set up an efficacy trial. Newly diagnosed patients with the appropriate markers were randomized between the five therapies across all the major centres that specialized in the disease. There was no need, nor ethical basis for a control group, since the natural course of the disease was well known and there was no valid argument for providing zero treatment to a group of ALS patients wishing to participate in the study.

The patients were followed by a mixture of physical performance markers and imaging studies in real time. It soon became apparent that two treatments held promise while others did not, and so the patients from the other three arms were randomized between the two promising therapies.

After 12 months further study of the two remaining therapies, the evidence on one of them was strong enough for the regulator to accept an application for conditional approval, for the subgroup identified by the specific markers. The second therapy was not as convincing: some patients had seen some disease progression but not at the rate typical for the disease and it was decided to run this arm of the trial for a further 12 months.

Monitoring of the patients prescribed the new treatment under conditional approval required their clinicians to report monthly on both disease-related performance status and blood markers found to be associated with the disease. Safety signals were also reported, and side-effects of the treatment, though quite serious,

Box 8.2 Continued

were judged as acceptable by the patients in relation to the disease being held at bay.

HTA agencies, involved in the initial design of the evidence generation plan, fast-tracked assessment based on the interim costing proposed by the company. They took into account the quality of life impact of the drug and also the impact on the family and the social care system of a substantial delay in the process of physical decline that ultimately required a wheelchair, 24-hour nursing, and other support services.

Since the patient numbers were small and sharply defined by the biomarkers, each national health system prepared to fund the new drug struck a fixed term 'reimbursement with evaluation' contract related to this volume. (Those countries that had co-funded the trials had pre-agreed discounts to the general terms.) Longer-term pricing would be established in a further 2 years on the basis of effectiveness in use of the expanded post-approval population.

So, within 4 years of the programme being initiated, conditional approval was given to the second treatment, on the basis of the 24-month study. It was found to be better tolerated than the first treatment but less efficacious. Despite the difference in efficacy, patients asked for the choice to be made available.

By the time 5 years had elapsed, the two effective treatments were being trialled in other sub-populations, since no additional drug candidates were yet ready for clinical trials. However, the in vitro and in vivo screens had been improved and several new compounds were showing promise.

The dreadful disease of ALS was finally in retreat.

References

1. **GenomeWeb** (2013) *OHSU, Intel to Collaborate on Cancer and Complex Disease Research*, https://www.genomeweb.com/informatics/ohsu-intel-collaborate-cancer-and-complex-disease-research, accessed 01 Sep. 2015.

2. **Fernandez E.** 2014. *New Breast Cancer Results Illustrate Promise and Potential of I-SPY 2 Trial: Trial Identifies Breast Cancer Patients Likely to Benefit from Experimental Drug*, https://www.ucsf.edu/news/2014/04/113151/new-breast-cancer-results-illustrate-promise-and-potential-i-spy-2-trial, accessed 01 Sep. 2015.

Index